Innovations in the Food System: Volume I

Innovations in the Food System: Volume I

Lyla Smith

Larsen & Keller
www.larsen-keller.com

Innovations in the Food System: Volume 1
Lyla Smith
ISBN: 979-8-88836-060-6 (Hardback)

Larsen & Keller
Published by Larsen and Keller Education,
5 Penn Plaza,
19th Floor,
New York, NY 10001, USA

Cataloging-in-Publication Data

Innovations in the food system: Volume 1 / Lyla Smith.
p. cm.
Includes bibliographical references and index.
ISBN 979-8-88836-060-6
1. Food supply. 2. Food industry and trade. 3. Food consumption. 4. Food.
5. Agricultural innovations. I. Smith, Lyla.
HD9000.5 .I56 2023
338.19--dc23

For more information regarding Larsen and Keller Education and its products, please visit the publisher's website www.larsen-keller.com

Table of Contents

Preface

The food system refers to the interconnected systems and processes that influence nutrition, food, health, community development and agriculture. It involves a series of activities including the production, processing, transport, distribution and consumption of food and related products. The recent innovations in agriculture have led to the development of sustainable agricultural practices. Sustainable food systems aim to provide healthy food to people along with creating food systems that are environmentally and economically sustainable. Conservation and yield-enhancing technological innovations including remote sensing, precision agriculture and no-tillage can lead to enhancement in productivity and efficient use of natural resources. The topics included in this book on innovations in the food system are of utmost significance and bound to provide incredible insights to readers. It will provide comprehensive knowledge to the readers.

The information shared in this book is based on empirical researches made by veterans in this field of study. The elaborative information provided in this book will help the readers further their scope of knowledge leading to advancements in this field.

Finally, I would like to thank my fellow researchers who gave constructive feedback and my family members who supported me at every step of my research.

Lyla Smith

Part I
Institutional Change, Scaling and RT&B Crops

1

Root, Tuber and Banana Crops: An Introduction

Graham Thiele, Michael Friedmann, Vivian Polar, and Hugo Campos

Abstract Root, tuber, and banana (RT&B) crops play a critical role in food and nutrition security in developing countries, increasingly so in sub-Saharan Africa (SSA). They have great potential to contribute to alleviate poverty, improve health and nutrition, and enhance the resilience of smallholder farmers to climate change. However, RT&Bs are characterized by unique challenges including vegetative propagation, genetic complexity, and postharvest constraints with bulkiness and perishability, compared to cereals. They are also characterized by a high yield potential and the ability to deliver micronutrients at large scale. However, until recently they have suffered from neglect in both investment and research. The CGIAR Research Program on Roots, Tubers and Bananas, which operated from 2012 to 2021, represented a novel and successful innovation model within the agricultural research for development domain not only in scientific terms but also from an organizational perspective. This program built upon the uniqueness of the RT&B crops and contributed to much of the progress reported in the book. This chapter provides an overview of challenges and opportunities facing RT&B crops in processing, marketing and distribution, enhancing productivity, and improving livelihoods. It presents underlying concepts for gender and scaling that feature prominently throughout the book, as well as an updated stance on innovation, touching base on the topic of the jobs to be done. The chapter concludes with an overview and highlights of the different sections and chapters in the remainder of the book.

G. Thiele (✉) · M. Friedmann · V. Polar
CGIAR Research Program on Roots, Tubers and Bananas (RTB), led by the International Potato Center, Lima, Peru
e-mail: g.thiele@cgiar.org; v.polar@cgiar.org

H. Campos
International Potato Center (CIP), Lima, Peru
e-mail: h.campos@cgiar.org

1.1 Introduction

Root, tuber, and banana (RT&B) crops play a critical role in food and nutrition security in developing countries, increasingly so in sub-Saharan Africa (SSA). They have great potential to contribute to alleviate poverty, improve health and nutrition, and enhance the resilience of smallholder farmers to climate change. However, RT&Bs are characterized by unique challenges and, until recently, have suffered from neglect in both investment and research (Fig. 1.1). The CGIAR Research Program on Roots, Tubers and Bananas (RTB hereafter), which operated from 2012 to 2021, represents a novel and successful innovation model within the agricultural research for development domain. This program built upon the uniqueness of RT&B crops, including vegetative propagation, genetic complexity, and postharvest constraints with bulkiness and perishability, compared to cereals, and also their high yield potential and the ability to deliver micronutrients at large scale.

Historically, RT&B crops have suffered from lower investment, and thus, this field of research features fewer researchers than other crops. In this context, RTB raised the profile of this important crop group for global food and nutrition security and served as an effective advocate for more investment. As different chapters in the book will make clear, RTB has not only focused on developing and implementing

Fig. 1.1 Value chains for bulky and perishable RT&B crops need upgrading. Market in Kampala, Uganda. (Photo credit: M. Friedmann (RTB))

particular innovations but has also addressed broader constraints for scaling innovations, since innovation only creates value when embraced and adopted by its users and beneficiaries at scale.

This book delivers an updated stance on innovations in RT&B crops, within and outside of the CGIAR program. We will touch on how to design innovations with a high likelihood of success and impact and provide several successful examples of innovations in RT&B crops while examining the scaling readiness of the innovations.

1.2 Importance of RT&B Crops

The importance of RT&B crops in developing countries has grown rapidly in the past 60 years: total production has quadrupled, and in this group, potato, cassava, and yam have experienced the highest growth. Sweetpotato declined a bit as producers in Asia who relied on sweetpotatoes as a food security crop and for animal feed switched to other crops (Table 1.1).

Table 1.1 Production and area harvested of RT&B crops and value of production in developing countries

# of countries	Crop	Production[a] (000,000 tonnes) 1961–1963	1988–1990	2016–2018	Area (000,000 ha) 1961–1963	1988–1990	2016–2018	Price USD/Mg[b]	Value USD billions	% of total
86	Cassava	74.6	150.0	281.1	9.8	15.0	24.7	128	36	10.7
90	Potato	26.8	76.3	221.5	3.5	6.3	12.1	318	71	20.9
96	Sweetpotato	93.1	121.0	88.5	12.7	9.1	7.9	741	66	19.5
101	Banana	21.1	54.4	112.9	2.0	4.1	5.5	621	70	20.8
63	Yam	5.9	18.7	72.9	1.2	2.0	8.5	743	54	16.1
46	Plantain	13.4	24.5	38.7	2.5	4.2	5.5	597	23	6.9
88	Other RT&B[c]	6.9	9.6	21.0	1.4	1.7	3.1	920	17	5.2
	Total	**241.8**	**454.5**	**836.6**	**33.1**	**42.4**	**67.3**		**337.0**	**100**

Source: Scott (2020)

[a]Production and area (FAOSTAT), 2020

[b]IFPRI's IMPACT model for commodity projections utilizes average prices (USD/Mg) in 2005 of 515.99 for banana and plantain, 673.1 for sweetpotato and yam, taro, and yautia, and 116.89 and 257.85 for cassava and potato, respectively. Estimated prices for 2030 are 735.61 for banana, 684.82 for plantain, 819.28 for yam, taro, and yautia, 815.26 for sweetpotato, 384.17 for potato, and 140.25 for cassava. Prices listed above are estimates for 2017 based on a linear interpolation of prices for 2005 and 2030 and then rounded to the nearest dollar. The prices for 2005 and 2030 are based on IFPRI researchers' determinations of past and future market tendencies as well as consultations with specialized scientists in international agricultural research centers around the world

[c]Other RT&B crops include the production (in million tonnes) of taro (10.2), yautia (0.5), and not specified R&T (10.3) for 2016–2018, with a production value (USD billions) of 7.6 for taro, 0.3 for yautia, and 9.9 for not specified R&T in 2016–2018

Table 1.2 Changing share of total cropped area in developing countries in Africa

	% area in Africa		
Crops	1961–1963	1989–1991	2016–2018
Banana	26.6	34.3	36.4
Cassava	57.1	57.7	75.5
Plantain	80.5	77.6	74.8
Potato	7.8	12.4	15.1
Sweetpotato	5.2	15.2	57.4
Yam	93.0	94.9	97.9

Source: FAOSTAT, New Food Balances. Last Update: February 19, 2020

Table 1.3 Contribution of RT&B crops to food intake in selected African countries in kilocalories (kcal) per capita (2017)

Country	Population (000s)	Grand total kcal	RT&B foods kcal	% RTBs foods
Democratic Republic of Congo[a]	62,523	1605	916	57
Ivory Coast	24,437	2730	970	36
Ghana	29,121	3033	1430	47
Nigeria	190,873	2464	622	25
Rwanda	11,981	2215	1046	47
Uganda	41,167	2144	659	31

[a]DRC 2008. Source: FAOSTAT, New Food Balances. Last Update: February 19, 2020

Rapidly increasing populations in Africa, growing conditions in the humid tropics, and a strong preference for RT&B crops mean that, with the exception of plantain, an increasing share of the area under these crops is grown in this continent (Table 1.2).

The contribution of foods derived from RT&B crops to calorific needs ranges from 25% in Nigeria to 57% in the Democratic Republic of Congo. Populations are growing rapidly in SSA, particularly in urban centers. By 2050, just these six countries (Table 1.3) are projected to have a combined population of 0.81 billion (UNDP 2019). The number of people involved in RT&B-dominated agri-food systems could more than double by the end of the twenty-first century—most of it in SSA.

Globally, RT&B crops are of particular importance to the poorest households in developing countries (Wiebe et al. 2020). They enhance the resilience of food systems and increase food availability and diversity, especially during hunger periods or in the event of crop failure or damage to cereals during extreme weather events. Through biofortification, RT&B crops can contribute to reducing widespread micronutrient deficiencies of vitamin A (VA), iron (Fe), and zinc (Zn), particularly in young children and women of reproductive age.

In Asia, yields of RT&B crops mostly showed large increases, doubling or tripling (Table 1.4; Fig. 1.2). In Latin America and the Caribbean (LAC), we see general increases but more modest. However, with the partial exception of potatoes and bananas, productivity of RT&B crops in SSA remains low. Hence, production does

Table 1.4 Change in yield for RT&B crops in developing countries

	Yield (t/ha)								
	Africa			Asia			LAC		
Crops	1961–1963	1989–1991	2017–2019	1961–1963	1989–1991	2017–2019	1961–1963	1989–1991	2017–2019
Bananas	6.1	6.9	11.0	8.7	15.1	31.7	16.4	18.6	24.6
Cassava	5.7	8.0	8.7	8.1	13.1	21.7	12.2	11.8	13.0
Plantains	4.7	5.2	6.0	6.6	9.7	12.5	8.1	8.6	11.0
Potatoes	7.9	10.8	14.8	8.7	12.7	20.1	7.0	12.4	19.7
Sweetpotatoes	5.2	5.2	6.3	7.5	15.4	19.6	8.1	7.0	10.9
Yams	7.2	9.7	8.5	10.3	12.2	14.5	7.0	7.3	8.0

Fig. 1.2 Harvesting improved cassava varieties in Dong Nai province, Vietnam. (Photo credit: G. Smith (Alliance of Bioversity & CIAT))

not satisfy basic food security needs in rural areas and does not compete well with imported staples for urban consumers, resulting in missed smallholder income opportunities. As people move to cities in SSA, value chains for RT&B crops need to be reconfigured to improve efficiency and convenience and reduce postharvest losses to compete with imports.

Outside of the humid tropics in Africa and in most of Asia and LAC, RT&B crops are generally important in rotations with cereals (Fig. 1.3) or legumes, as part of agroforestry systems, such as bananas and coffee, or as secondary crops (Fig. 1.4). In this way, they contribute resilience, increase efficiency of cropping systems, provide key micronutrients in the diet, and generate off-season income.

Fig. 1.3 The Assam Agribusiness and Rural Transformation Project (APART) project in Assam State, India, introduced potato farmers to climate resilient zero tillage with paddy straw mulch potato production. (Photo credit: CIP)

Fig. 1.4 Mr. John Ndamira in his coffee/banana plantation, with I. Arinaitwe of Mbarara District, Bushenyi, Uganda. (Photo credit: M. Friedmann (RTB))

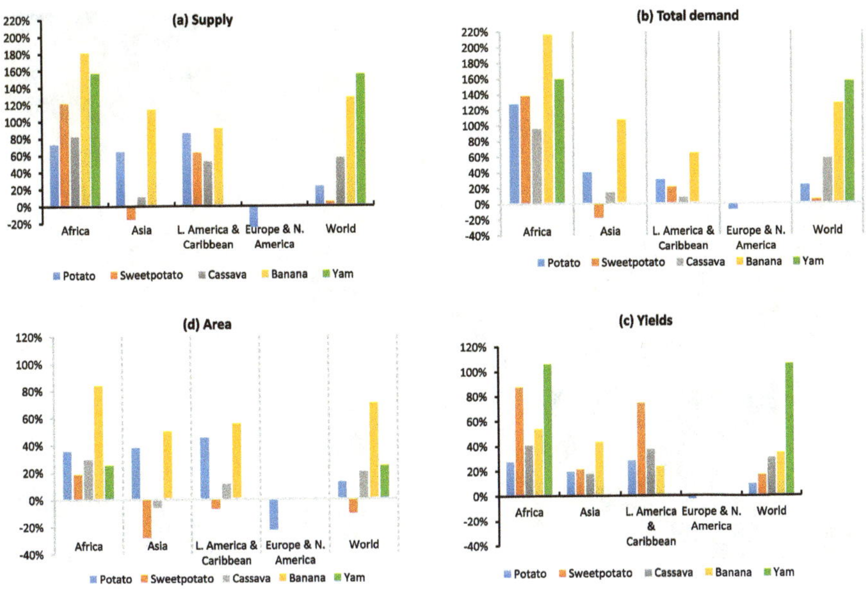

Fig. 1.5 Changes in RT&B agriculture from 2010 to 2050 under a baseline foresight scenario. (Source: Authors' calculations based on Rosegrant et al. 2017)

Using the global partial equilibrium model IMPACT, future agricultural demand and supply for RT&B crops were projected, considering future climate and socio-economic changes. These forecasts (Fig. 1.5) predicted RT&B crop production reaching 1400 million tons (of fresh produce) by 2050—a global increase of almost 50% from 2010—with Africa emerging as the world's largest producing and consuming region (Petsakos et al. 2019).

1.3 Challenges and Opportunities for RT&B Crops

Significant technological change will be needed to cope with climate change, especially in SSA, where RT&B crops are of the highest importance (Thiele et al. 2017). Therefore, a key dynamic will be crop substitution, especially in areas where RT&B crops can replace more climate-sensitive cereals and legumes. Maize is vulnerable to higher frequencies of periodic drought, whereas production of cassava and sweetpotato can be much more reliable under these conditions. In addition, there are specific traits that make RT&B crops tolerant or resistant to abiotic stresses like heat, drought, soil salinity, and waterlogging and shocks like typhoons/cyclones (Fig. 1.6). For this reason, roots and tubers are important contributors to post-disaster recovery and mitigation for their capacity for piecemeal harvesting, underground protection, and short growing cycles (Prain and Naziri 2020). Sweetpotato has played an especially important role in humanitarian relief operations combining rusticity, short crop cycles and—particularly with orange-fleshed varieties—high levels of micronutrients (Fig. 1.7).

Fig. 1.6 A field affected by typhoon with broken trees and barely damaged sweetpotato plants, Luzon, Philippines. (Photo credit: CIP)

Fig. 1.7 High β-carotene (pro-vit A) OFSP. (Photo credit: D. Gemenet (EiB, CIP))

1.3.1 Processing, Marketing, and Distribution

Because RT&B crops are more perishable and bulkier than cereals, they create opportunities for value addition and employment in postharvest and processing in rural areas, especially for women. Providing growing urban populations with

Fig. 1.8 A mechanical sieve—coarse mash retained in the wooden box and fine mash captured in the basin after separation. (Photo credit: I. Okike (IITA))

RT&B crop-based food will require extensive transformation of current technology to capture these benefits. But unless gender roles and needs are systematically considered, innovation can exacerbate gender inequality (Sarapura 2012). Increasing opportunities for women can have a powerful impact on productivity and agriculture-led development and help to reduce gender gaps in access to inputs, assets, opportunities, information, and other resources (Margolies and Buckingham 2013; FAO 2014).

RT&B crops and their residues are finding increasing use in animal feed. High-quality cassava peels (HQCP; Fig. 1.8) are being taken to scale in Nigeria (with interest elsewhere in SSA), while sweetpotato silage is now widely used in Asia and finding greater popularity in Uganda (Asindu et al. 2020).

Since cassava starch is particularly important for its specific functional properties and constitutes the largest source of starch in tropical regions, improving small- and medium-scale processing technologies is key to expanding the market in SSA (Chapuis et al. 2016).

1.3.2 Enhancing Productivity

Investment in RT&B crops compared to cereals has lagged, slowing yield growth. RT&B crops offer high potential yields, but farmers often realize less than half the potential due to poor quality planting material of limited genetic potential, biotic and abiotic constraints, and poor management practices. Seed systems may not work well because of inappropriate policies for RT&B crops or where famine and disaster-relief undermine the development of healthy seed market systems by providing large volumes of free propagation materials. These bottlenecks in markets and policy further restrain the willingness to invest in yield-increasing input intensification. Where market conditions are favorable and appropriate new technology available, yield gains have been considerable, as in Southeast Asia for cassava (Malik et al. 2020). In Africa, potato is grown increasingly as a cash crop for urban markets, showing consistent growth with the second highest rate (after banana) of yield increase of any crop over the last decade.

Recent adoption studies in SSA provide optimism that yield gaps can be closed. For instance, in Nigeria, modern varieties of cassava are grown on 39.9% of cultivated area with a yield increase of at least 60%, bringing an estimated 1.6 million people out of poverty each year (Wossen et al. 2018). However, in general, adoption rates of modern varieties of RT&B crops are below 40%; therefore, more investment in breeding and seed systems is required (Thiele et al. 2020). Seed systems for RT&B crops can be a particular challenge, as farmers can more easily share and retain planting material that discourages the interest of private seed companies. Hence, some public sector seed system investment is usually required.

Women's farm yields and incomes are typically much lower than men's, reflecting specific gender-related barriers that affect women's productivity: competing priorities (e.g., childcare), exclusion from farming decisions by men, and limited access to land, markets, and information technology (FAO 2014; Mudege et al. 2015). COVID-19 may be undoing recent progress on reducing gender gaps and compromising food and nutrition security (Doss et al. 2020).

RTB has paid explicit attention to gender in its program, and gender is a specific focus of this book, which uses the following definitions of key terms for these important discussions (Box 1.1).

1.3.3 Improving Livelihoods

With an average production of approximately 820 million tons on 64 million hectares (ha) in 2016–2018 (FAOSTAT 2020), RT&B crops represent the second most important set of crops in developing countries after cereals. The yield potential of RT&B crops is very high, providing one of the cheapest and more affordable sources of dietary energy (Lebot 2020). In 2017, RT&B crops alone provided around 10% of the daily per capita calorie intake for the 864 million people living in least developed countries (Kennedy et al. 2019).

RT&B crops can also play an important role in nutrition security. In SSA, vitamin A deficiency (VAD) is widespread, contributing to increased risks of blindness, illness, and premature death, particularly in young children and pregnant/postpartum women. Globally, 163 million children under 5 years of age are vitamin A deficient, and prevalence rates of Fe and Zn deficiencies are even higher. Orange-fleshed sweetpotato (OFSP) is a proven biofortified crop: 100 g/day can meet the vitamin A requirements of a young child (Fig. 1.9), and 6.2 million households

Box 1.1 Key Terms for Discussing Gender

Gender is not the same as sexuality. Gender is socially constructed and refers to roles and behaviors of men and women that are the intended or unintended products of social practice.

Gender analysis is a critical examination of how differences in gender roles, activities, needs, opportunities, and rights affect women, men, girls, and boys in a given situation or context.

Gender gap is the difference between women and men in terms of their levels of participation, access, rights, remuneration, or benefits.

Gender inequality is reflected in unequal access to or enjoyment of rights, remuneration, or benefits, as well as the assumption of stereotypical social and cultural roles.

Gender roles and norms are standards or expectations of the behavior of women and men, just because they are male or female. These expectations are particular for each society, culture, and community at a given point in time.

Gender relations are social relations that unite women and men as social groups in a given context, including how power and access to and/or control over resources are distributed between the sexes.

Gender power relations are how gender shapes the distribution of power at all levels of society.

Gender responsive describes the quality of a specific tool, approach, or intervention in terms of actively examining and addressing gender norms, roles, and inequalities. This type of approach often involves actions to create an environment that promotes gender equality.

Gender transformative describes the quality of a specific tool, approach, or intervention in terms of creating opportunities for individuals to actively challenge and/or transform gender norms to address power inequalities between people of different genders.

Gender intentional describes the quality of a specific program, project, or intervention in terms of purposefully addressing gender considerations in the design and/or operation. It seeks to target and benefit a specific gendered group to achieve specific goals.

Intersectionality is an analytical tool for studying, understanding, and responding to the ways that sex and gender intersect with other personal characteristics or identities and how these intersections contribute to unique experiences of discrimination.

Empowerment is the process by which individuals gain power and control over their own lives and acquire the ability to make strategic choices.

Sex-disaggregated data is data collected and tabulated separately for women and men allowing the measurement of differences between women and men in terms of various social and economic dimensions.

Fig. 1.9 Mothers and caregivers learning how to prepare dishes including orange-fleshed sweetpotato safely while retaining vitamins to improve their young children's nutrition (Photo credit: H. Rutherford, CIP)

(HH) have been reached with improved sweetpotato varieties—mostly OFSP—across 15 African countries (Low and Thiele 2020). HarvestPlus and its partners delivered cassava varieties enriched with vitamin A to more than one million farming households in Nigeria and the Democratic Republic of Congo (Ilona et al. 2017). Banana cultivars are also a significant source of vitamin A (Amah et al. 2018) and are being promoted in East Africa. Meanwhile, potato breeding has achieved significantly enhanced levels of Fe and Zn (Amoros et al. 2020). Where consumed as a staple, biofortified potatoes can contribute up to 50% of women's requirements for these micronutrients (Burgos et al. 2019). Jongstra et al. (2020) have recently found that Fe from biofortified potatoes is more bioavailable to humans than any other Fe biofortified crop developed to date.

Much current work on RT&B crops in SSA and Asia is demonstrating the power and effectiveness of these crops in alleviating poverty, fighting hunger and nutrition insecurity, especially within the current content of climate change and COVID-19 (Heck et al. 2020). RT&B crops are mostly produced, processed, and traded locally, making them less vulnerable to abrupt price fluctuations in international markets and interruptions due to epidemics. RT&B-based products (such as orange-fleshed sweetpotato puree used in baked goods) have been developed and promoted and are currently going to scale in African markets. There is growing global demand for innovative ingredients to meet growing consumer tastes and nutrition needs. Their potential is often limited, however, by the lack of preferred nutritious varieties, instability of micronutrients particularly when processed, unfavorable value chains, and weak institutional arrangements.

1.4 A Primer on Innovation and the Jobs to Be Done

It is clear that for RT&B crops to contribute more extensively, innovation is required in all the domains so far described. The RTB program put innovation and the scaling of innovations at center stage.

As part of its program design, RTB invested in a subprogram—known as a "flagship project" in CGIAR—dedicated to improving livelihoods by scaling RT&B solutions in agri-food systems. The flagship provided support to next and end users for scaling of RTB innovations from the farm to the community, the region, and beyond. We guided scientists in the program on how to better target research, design strategic youth and gender research, select partners, and develop capacity for improving livelihoods at scale. This results-oriented approach included the following:

1. Forward-looking analysis of trends
2. Decision support for the tailoring, integration, and scaling of RT&B technologies based on farmer typologies and practical investment steps
3. Identifying best possible RT&B (and other crops and livestock) technology options that contribute to sustainable livelihoods
4. Addressing constraints for RT&B technology adoption with gender and intergenerational transformations
5. Evidence base to improve scaling of RT&B agri-food system innovations with enhanced equity

The flagship developed an approach to innovation and scaling that was widely applied within the program through a dedicated fund for scaling (see Chap. 2) and is featured in several chapters of this book as a key crosscutting topic. Key definitions for innovation and scaling are provided in Box 1.2.

Several chapters in this book use the concept of Scaling Readiness to characterize different innovation packages (Fig. 1.10). "Innovation readiness" refers to the demonstrated capacity of an innovation to fulfill its contribution to development outcomes in specific locations. This concept is presented in nine stages showing progress from an untested idea to a fully mature and proven innovation. "Innovation use" indicates the level of use of an innovation or innovation package by the project members, partners, and society. This concept captures progressively broader levels

Box 1.2 Key Definitions for Innovation and Scaling

Innovations are the new ideas, products, services, and solutions capable of facilitating impact through innovation systems involving multiple partners and enablers.

Innovation systems are the interlinked sets of people, processes, assets, and social institutions that enable the introduction and scaling of new ideas, products, services, and solutions capable of facilitating impact.

Scaling of innovations is a deliberate and planned effort to enable the use of innovations to have positive impact for many people across broad geographies.

Impact is a durable change in the condition of people and their environment brought about by a chain of events to which research, innovations, and related activities have contributed.

Scaling strategy is a set of coherent activities, stakeholders, and stakeholder engagement models to enable scaling.

Innovation package is the combination of innovations that are needed for scaling in a specific location or context.

Scaling approach is an integrated set of scaling tools and procedures that can be used to design and implement scaling activities in different contexts.

Innovation readiness refers to the demonstrated capacity of an innovation to fulfill its contribution to development outcomes in specific locations. This is presented in nine stages showing progress from an untested idea to a fully mature proven innovation.

Innovation use indicates the level of use of the innovation or innovation package by the project members, partners, and society. This shows progressively broader levels of use beginning with the intervention team who develops the innovation to its widespread use by users who are completely unconnected with the team or their partners.

Scaling Readiness of an innovation is a function of innovation readiness and innovation use. Table 1.5 provides summary definitions for each level of readiness and use adapted from Sartas et al. (2020), which have been used throughout the book. Scaling Readiness also is the name of the approach to scaling described in Chap. 3.

Table 1.5 *Definition* of levels of innovation readiness and use (Sartas et al. 2020)

Stage	Innovation readiness	Innovation use
1	Idea	Intervention team
2	Basic model (testing)	Direct partners (rare)
3	Basic model (proven)	Direct partners (common)
4	Application model (testing)	Secondary partners (rare)
5	Application model (proven)	Secondary partners (common)
6	Application (testing)	Unconnected developers (rare)
7	Application (proven)	Unconnected developers (common)
8	Innovation (testing)	Unconnected users (rare)
9	Innovation (proven)	Unconnected users (common)

Responsible scaling requires ethics of co-responsibility for ensuring that the impacts from the innovation are well captured by the intended beneficiaries and minimize negative social consequences, whether these impacts are intentional or not, and whether they can be fully foreseen or not. One crucial dimension of responsible scaling is gender equity, which can be achieved through gender-responsive research. This idea was explicitly included in much RTB research and is featured in several chapters of this book as a cross-cutting topic (see Box 1.1 for definitions used in the RTB program and this book).

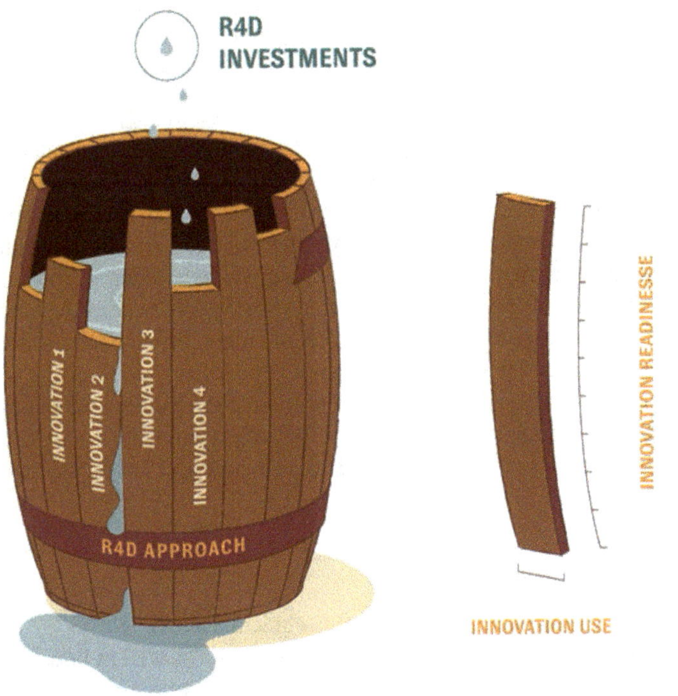

Fig. 1.10 Scaling Readiness barrel to illustrate how innovation(s) with the lowest readiness limits an innovation package's capacity to achieve impact at scale. (Adapted from "von Liebig's barrel," after Whitson and Walster, 1912; published in Sartas et al. 2020)

of use beginning with the intervention team who develops the innovation to its widespread implementation by users who are completely unconnected with the team or their partners. "Scaling Readiness" of an innovation is a function of innovation readiness and innovation use. Table 1.5 provides summary definitions for each level of readiness and use adapted from Sartas et al. (2020).

Many innovations fall short in adoption and value creation because they lack understanding about what drives a user/beneficiary/customer to choose one good or service over another. The "jobs to be done" (JTBD hereafter) theory, developed by the late Clayton M. Christensen and colleagues at Harvard University, provides a compelling perspective about why people buy or decline what companies try to sell them, or why beneficiaries from organizations involved in international development sometimes refuse to adopt what is offered or delivered. The JTBD theory significantly increases the ability to predict the likelihood of innovation success, which is a welcome addition to the toolbox of innovators. Its main tenet is that users do not purchase goods or services. Instead, people bring things into their lives to do a specific job or to achieve progress toward a particular goal under specific circumstances. Therefore, the goods and services that users acquire are unconscious means for filling satisfaction gaps. Unless we understand the "jobs" that people want to fill, the likelihood of innovation success is severely undermined, regardless of budget size or innovator skill. Every "job" exists within a given context—the where, when, who, or what of a situation. As just one example, a female smallholder in SSA tries to provide healthy, affordable food to her family. How can RT&B crops meet that need for her?

JTBDs can be quite different from solutions, and it helps to understand why well-funded innovation efforts run by very talented people oftentimes fall short of expectations and adoption, or why their scaling up stalls. For instance, many providers of MP3 reproducers focused on the presumed solution, namely, providing music. In contrast, Apple aimed the iPod at helping customers listen to music through a seamless experience. They reconsidered the whole business around personal music management, enabling customers to acquire, organize, listen to, and share music. In addition, Apple considered social and emotional factors involved in why people want to listen to music on the go and share the experience with others. Despite being a late entrant, Apple took over the market, beating formidable competitors such as Sony and Windows, both of which had more resources, larger teams, and much more cash available.

By far, the most valuable insight from the JTBD theory is that any job has not one but rather three components:

- *Functional*, which addresses practical, technical aspects
- *Social*, which addresses behavioral aspects of users, as well as how the users take into account the effects that the action will have on those around them
- *Emotional*, which relates to deep, personal aspects of users such as feelings, sentiments, aspirations, and desires

Regrettably, since most innovation efforts are driven by their functional components from their onsets, their likelihood of success is in jeopardy since the social and emotional components are the ones that represent a strong, if not the strongest, driver of user preferences and subsequent adoption. Within the agricultural domain, innovations solely focused on productivity or technological progress but which fail to account for cultural and social aspects are equally in jeopardy. More detailed accounts about the JTBD theory and its use in agricultural contexts can be found in Christensen et al. (2016) and Campos (2021).

1.5 Layout of the Book and Key Lessons to Enable Effective Innovation in RT&B Seed Systems

The body of this book is organized by four broad topics:

1. Institutional change and scaling
2. Processing and marketing
3. Productivity
4. Improving livelihoods

In each of these sections, we identify key lessons for designing and enabling effective innovation efforts in RT&B crops.

1.5.1 Institutional Change and Scaling

In the first section on institutional change and scaling, Chap. 2 looks at the constituent institutional innovations that underpinned the RTB program and how they added value to the combined achievements of the participating centers and their partners in science and research for development outcomes (which are described in more detail in the consequent chapters). The 10-year RTB program raised the profile of RT&B crops with investors within the research and development community. Dedicated funding expanded the frontier of knowledge around several of the key constraints to expanded utilization intrinsic to vegetatively propagated crops, particularly in the areas of breeding, seed systems, pests and diseases, and postharvest utilization. Within the domain of breeding, RTB excelled at providing tools and knowledge for breeders to acquire the skills necessary to develop product profiles (a set of traits and specifications to inform breeding efforts) in such a way that improved varieties are released with a greater likelihood of adoption by smallholder farmers.

Progress by RTB was facilitated by an array of institutional innovations for collaborative governance and management, including stakeholder consultations and priority setting, a portfolio organized by aggregated innovations (i.e., clusters of activities), articulated flagship projects, incentive funds, a dynamic interactive communication capacity, programmatic embedding of strategic and integrated gender research, and purposive engagement of national partners. Overall, RTB's design, governance, and management innovations added value to the combined achievements of the participating centers in science and research for development outcomes that are described in many of the book chapters.

Another key learning evident from the book is that there were limitations in the degree to which these institutional innovations could be "scaled down," from the program level to the implementation of projects and activities. This limitation was, in part, a reflection of the predominance of bilateral funding with multiple donors under center management, which made "scaling down" outside of earmarked funding/scaling projects challenging.

RTB also pioneered support and investment in the development and implementation of scaling strategies and partnerships to catalyze the scaling of its innovations. This body of work and linked research supported the development of the Scaling Readiness approach, which is widely relevant beyond RT&B crops. Chapter 3 covers the principles and practices of Scaling Readiness, which assesses the readiness of innovations to be scaled and supports the development, implementation, and monitoring of scaling strategies at project and organizational levels. The chapter explores how the approach enabled different RT&B innovations to move to scale. The innovations described in Chaps. 4, 5, 6, 10, and 12 are described in relation to Scaling Readiness and benefited directly from RTB investments in a Scaling Fund and the application of the Scaling Readiness approach. Chapter 17, meanwhile, used the Scaling Readiness framework to assess progress in scaling biofortification across multiple RT&B crops. A key lesson from this work is that a focused scaling approach adds value particularly when dedicated funding is available. However, these investments are effective only if tested guidelines and capacity development with specialized roles for scaling in the team are also available.

1.5.2 Processing, Marketing, and Distribution

The second section addresses innovations in processing, marketing, and distribution of RT&B crop products. The RTB program had a dedicated flagship for adding value to the processing of RT&B crops and enhancing value chains, with targeted investments through earmarked funding and scaling fund grants (as described in Chap. 2). This idea was built on the insight that RT&B value chains could make more nutritious food more cheaply and widely available in cities and rural markets by improving fresh supply, by developing new RT&B food products, and by supporting different typologies of small- and medium-scale processors.

Chapter 4 examines the scaling of innovations to improve the overall efficiency of small-scale cassava industries for rural small-scale processors to access more distant and higher-value markets (Fig. 1.11). Chapter 5 investigates sweetpotato puree as a novel ingredient in the food industry, developed as an innovative value chain product to address many constraints of sweetpotato adoption and use, while delivering a highly nutritional ingredient for processed bakery goods for rural and urban populations. Together, these chapters show how investments in cassava and sweetpotato processing addressed the challenges of limited shelf life for these two crops and how they overcame bottlenecks in their conversion to value-added products.

Fig. 1.11 Toasting grated and fermented cassava is the final step in creating gari and is a task mainly done by women. (Photo credit: H. Holmes (RTB))

In the first case, improving the energy efficiency of drying cassava into flour and starch through the development and scaling out of flash dryers led to 10–15% gains in productivity and incomes for small- and medium-scale processors. In the second case, development of orange-fleshed sweetpotato puree and developments in the value chains from field to market enabled bakeries in Ghana, Kenya, Malawi, and Rwanda to substitute puree for 20–50% of the wheat flour used in cookies, donuts, and bread by some commercial bakeries. By reaching urban markets, demand for OFSP was created and nutritional outcomes improved as a result. Between the two cases, we note that cassava processors could use locally manufactured equipment while OFSP puree required imported equipment. This difference has implications for the scaling strategies and long-term sustainability—two topics that merit further analysis.

Chapter 6 looks at the conversion of waste cassava peels into a valuable animal feed ingredient while also addressing serious environmental issues created by the millions of tons of waste cassava peels generated during processing. This chapter shows how it is possible to build on existing technologies and capacities to add value by processing waste cassava peels into animal feed. However, various "soft" components of scaling needed attention here as they formed the main bottlenecks to expanding the processing of high-quality cassava peels (HQCP).

Chapters 7 and 8 look beyond the innovations in cassava of the RTB program by focusing on cassava-related technology transfer between Brazil and Africa. Chapter 7 shows that for existing and well-developed processing technologies to be transferred to potential new users, the capacity, motivation, and training of the targeted users will determine the uptake and success. Chapter 8 notes the need for renewed attention on food safety needs as processing acquires greater importance in developing countries, especially countries in Africa. While this group of chapters record technological progress, they also emphasize the need for research on processing innovations to focus on women and men operators and entrepreneurs and to avoid getting lost in the details of machinery, products, and processes. In other words, RT&B crops require more study along this theme with attention to gender in the design, adaptation, and scaling of processing technologies.

1.5.3 Enhancing Productivity

The third section addresses innovations to enhance productivity, especially around pest and disease management and the challenges of seed systems for vegetatively propagated crops.

Pests and diseases are major production constraints in vegetatively propagated crops such as RT&Bs. Chapter 9 documents how digital technologies have been developed to identify and monitor emerging pests and diseases, looking at their future potential but also the challenges, limitations, and innovative approaches taken to reach end users, particularly smallholder farmers. This work illustrates that cohesive engagement between digital tool developers, researchers, extension and advisory services, and farmers is required to define specific problems and opportunities for developing suitable digital solutions. Moreover, the use of digital tools is still in its infancy with RT&B cropping systems; thus, opportunities are available for innovation for RT&B farming systems.

Chapter 10 looks at the process, practices, challenges, lessons learned, and future policy implications associated with scaling banana Xanthomonas bacterial wilt (BXW) management practices—an endeavor that has enabled the recovery of substantial banana production areas. This work provides evidence of the effectiveness of single diseased stem removal (SDSR) to manage and control BXW (Fig. 1.12). Nevertheless, the efforts to reduce the labor intensity of the SDSR package are equally important. Action plans for training and information dissemination related to SDSR must consider gender aspects at the onset of interventions if maximum participation is to be reached with women and youths. In particular, the SDSR case in Rwanda suggests the importance of engaging policy makers at all stages of technology innovation to secure their buy-in for subsequent scaling to the intended users.

Chapters 11,12, 13, 14, and 15 expand on the work to strengthen and enable seed systems for increased access to improved varieties and for encouraging improved quality in places where informal seed systems still predominate. As seed multiplication and dissemination for RT&B crops face common and difficult constraints, RTB developed a toolkit to understand, assess, and support the design of VPC seed system interventions to improve their effectiveness. Chapter 11 reviews researcher experiences in using these tools among different crops and in different regions. Their reflections provide insights on how the toolbox has influenced our understanding of seed system performance. Work with the RTB toolbox also shows how its success depended on attracting and retaining the interest of public and private

Fig. 1.12 Application of the novel SDSR technique to control BXW: a woman in eastern DR Congo cuts a diseased stem at soil level. (Photo Credit: B. Van Schagen (Alliance Bioversity-CIAT))

sector actors to diagnose and improve VPC seed systems. Assembling the tools in one toolbox has made them more accessible, provided an intuitive structure for uninitiated users, and helped to clarify which tools and combinations of tools are most useful for addressing different types of challenges.

Chapter 12 looks at the Triple S (*S*torage in *S*and and *S*prouting) strategy to conserve and multiply sweetpotato planting material during long dry seasons, exploring evidence from Ethiopia and Ghana regarding the various communication channels used to influence the uptake of Triple S among male and female farmers. Chapter 13 examines the development and implementation of new early generation potato seed technologies, such as aeroponics and apical cuttings in East Africa, and their contribution to the dissemination of improved potato varieties. The transformation of yam seed systems is covered in Chap. 14, where the scaling of high-quality seed production technologies in seed companies in partnership with national regulatory agencies is delivering basic and certified seed yam tubers to seed entrepreneurs—a chain of activity that has resulted in localized, scalable deployment centers. The development of commercially sustainable cassava seed systems in the vast informal seed systems of Nigeria and Tanzania is explored in Chap. 15, with key lessons drawn for the further development of commercial cassava seed systems in Africa (Fig. 1.13).

Fig. 1.13 A village seed entrepreneur sells bundles of cassava stems as part of a market promotion by the BASICS program in Benue State, Nigeria (CRS)

Across the different seed system innovations, it became evident that partners have different capacities and capabilities that determine their role in contributing to the scaling process. In addition, several enabling factors related to policy, infrastructure, mindsets and behavior change, and marketing and value chains are critical for scaling innovation packages. Availability of quality seed and opportunities for commercialization were shown to have important gender implications. Continued work to understand and design gender-responsive communication channels in different sociocultural contexts is required.

Chapter 16 closes the section by exploring how to build demand-led and gender-responsive breeding programs and looks at the piloting of practical tools that enable plant breeders to examine gender differences in the targeted population of users and with regard to trait prioritization (Fig. 1.14). This chapter underscores the importance of having a good foundation for gender-responsive breeding, created by cultivating gender awareness though cross-disciplinary dialogue, gender training targeted to breeding programs, and the incorporation of a gender specialist into breeding teams.

1.5.4 Improving Livelihoods

The fourth and final section looks at improving livelihoods with biofortified RT&B crops playing a major role as an affordable source of key micronutrients in an emerging and more resilient food system in SSA. Chapter 17 examines the scaling and impacts of biofortified RT&B crops and how the lessons learned from these efforts can help illuminate pathways for having biofortified RT&B crops contribute to building back a better, climate-smart, and more nutritious food system in the

Fig. 1.14 Poundability is an important trait for West African women who pound boiled yam for traditional dishes. Portioning pounded yam, Bouaké, Côte d'Ivoire. (Photo credit: D. Dufour (CIRAD))

post-pandemic world. RT&B crops are well-positioned to move forward in the context of emergency recovery as well as gender-responsive food system transformation. As scaling efforts expand, managing the perishability and the seasonality of RT&B crops at larger scales will require greater investment in physical market chain infrastructure, storage, information systems, and enterprise development. Moreover, given the poor families' heavy dependence on staple foods, increased commitment to breeding for enhanced micronutrient and protein content is warranted (Fig. 1.15). Advocacy by the Scaling Up Nutrition movement and the recognition of biofortification by different African governments are two examples of the kind of policy engagement needed to keep biofortification and nutrition at the forefront of food policy and investment planning.

1.6 What Next for RT&B Crops and the Smallholder Farmers Relying on Them

We are writing this book during unprecedented change in the CGIAR at a time when the COVID-19 pandemic is deeply, and quickly, changing the way communications and work are done. The RTB program that generated much of the research reported

Fig. 1.15 Potatoes that are rich in iron can help alleviate the suffering of women and children who suffer from anemia. Bertha Azursa Clemente and her son with biofortified potatoes in Huancavelica, Peru. (Photo credit: S. Fajardo (CIP))

here will be transitioning into a new set of collaborative arrangements under the One CGIAR. So the book is both a reflection of what the program achieved and also a contribution to future research endeavors. We are optimistic that the increased attention, research investment, and innovations achieved bode well for the future, ultimately enhancing the livelihoods of the many millions of women and men involved in their production, processing, and consumption in RT&B food systems in developing countries.

References

Amah D, van Biljon A, Brown A, Perkins-Veazie P, Swennen R, Labuschagne M (2018) Recent advances in banana (Musa spp.) biofortification to alleviate vitamin A deficiency. Crit Rev Food Sci Nutr 59(21):1–39. https://doi.org/10.1080/10408398.2018.1495175

Amoros W, Salas E, Hualla V, Burgos G, De Boeck B, Eyzaguirre R, Zum Felde T, Bonierbale M (2020) Heritability and genetic gains for iron and zinc concentration in diploid potato. Crop Sci 60:1884–1896. https://doi.org/10.1002/csc2.20170

Asindu M, Ouma E, Elepu G, Naziri D (2020) Farmer demand and willingness-to-pay for sweetpotato silage-based diet as pig feed in Uganda. Sustainability 12(16):6452. https://doi.org/10.3390/su12166452

Burgos G, Zum Felde T, Andre C, Kubow S (2019) The potato and its contribution to human diet. In: Campos H, Ortiz O (eds) The potato crop. Springer International Publishing, Cham, pp 37–74. https://doi.org/10.1007/978-3-030-28683-5

Campos H (2021) The innovation revolution in agriculture. Springer International Publishing, Cham, 234 p. https://doi.org/10.1007/978-3-030-50991-0

CGIAR (2020) Scaling Brief #4: scaling glossary. Deutsche Gesellschaft für Internationale Zusammenarbeit, Bonn. https://hdl.handle.net/10568/110632

Chapuis A, Precoppe M, Méot J, Sriroth K, Tran T (2016) Pneumatic drying of cassava starch: numerical analysis and guidelines for the design of efficient small-scale dryers. Dry Technol. https://doi.org/10.1080/07373937.2016.1177537

Christensen CM, Hal T, Dillon K, Duncan DS (2016) Know your customers' "jobs to be done". Harv Bus Rev 96:374–382

Doss C, Njuki J, Mika H (2020) The potential intersections of Covid-19, gender and food security in Africa. AgriGender: J Gend Agric Food Secur 5(1):41–48. https://doi.org/10.19268/JGAFS.512020.4

FAO (2014) Innovation in family farming. FAO, Rome. http://www.fao.org/publications/sofa/2014/en/

FAOSTAT (2020) Crops database. http://www.fao.org/faostat/en/#data/FBS. Accessed 22 Dec 2020

Heck S, Campos H, Barker I, Okello JJ, Baral A, Boy E, Brown L, Birol E (2020) Resilient agri-food systems for nutrition amidst COVID-19: evidence and lessons from food-based approaches to overcome micronutrient deficiency and rebuild livelihoods after crises. Food Secur 12:823–830. https://doi.org/10.1007/s12571-020-01067-2

Ilona P, Bouis HE, Palenberg M, Moursi M, Oparinde A (2017) Vitamin A cassava in Nigeria: crop development and delivery. Afr J Food Agric Nutr Dev 17(2):12000–12025. https://www.ajfand.net/Volume17/No2/Biofortification%20Issue%20Chapter%209.pdf

Jongstra R, Mwangi MN, Burgos G, Zeder C, Low JW, Mzembe G, Liria R, Penny M, Andrade MI, Fairweather-Tait S, Zum Felde T, Campos H, Phiri KS, Zimmermann MB, Wegmüller R (2020) Iron absorption from iron-biofortified sweetpotato is higher than regular sweetpotato in Malawian women while iron absorption from regular and iron-biofortified potatoes is high in Peruvian women. J Nutr 150(12):3094–3102. https://doi.org/10.1093/jn/nxaa267

Kennedy G, Raneri J, Stoian D, Attwood S, Burgos G, Ceballos H, Ekesa B, Johnson V, Low JW, Talsma EF (2019) Roots, tubers and bananas: contributions to food security. In: Encyclopedia of food security and sustainability, vol 3. Elsevier, Oxford, pp 231–256. https://doi.org/10.1016/B978-0-08-100596-5.21537-0

Lebot V (2020) Tropical root and tuber crops: cassava, sweet potato, yams and aroids. CABI, United Kingdom. https://www.cabi.org/bookshop/book/9781789243369/

Low JW, Thiele G (2020) Understanding innovation: the development and scaling of orange-fleshed sweetpotato in major African food systems. Agric Syst 179(102770):15. https://doi.org/10.1016/j.agsy.2019.102770

Malik AI, Kongsil P, Nguyễn VA, Ou W, Srean P, Becerra López-Lavalle LA et al (2020) Cassava breeding and agronomy in Asia: 50 years of history and future directions. Breed Sci 70(2):145–166. https://doi.org/10.1270/jsbbs.18180

Margolies A, Buckingham E (2013) The importance of gender in linking agriculture to sustained nutritional outcomes. In: Agriculture and Nutrition Global Learning and Evidence Exchange (AgN-GLEE), Joint USAID & SPRING conference, Guatemala City, Guatemala from 5–7 March. https://www.spring-nutrition.org/sites/default/files/2.5-a_the_importance_of_gender_amy_margolies_elizabeth_buckingham.pdf

Mudege NN, Chevo T, Nyekanyeka T, Kapalasa E, Demo P (2015) Gender norms and access to extension services and training among potato farmers in Dedza and Ntcheu in Malawi. J Agric Educ Ext 22(3):291–305. https://doi.org/10.1080/1389224X.2015.1038282

Petsakos A, Prager SD, Gonzalez CE, Gama AC, Sulser TB, Gbegbelegbe S, Kikulwe EM, Hareau G (2019) Understanding the consequences of changes in the production frontiers for roots, tubers and bananas. Glob Food Sec 20:180–188. https://doi.org/10.1016/j.gfs.2018.12.005

Prain G, Naziri D (2020) The role of root and tuber crops in strengthening agri-food system resilience in Asia. A literature review and selective stakeholder assessment. International Potato Center, Lima. isbn:978-92-9060-539-3, 66 p. https://cgspace.cgiar.org/handle/10568/106669

Rosegrant MW, Kulakow P, Mason-D'Croz D, Cenacchi N, Nin-Pratt A, Dunston S, Zhu T, Ringler C, Wiebe K, Robinson S, Willenbockel D, Xie H, Known H-Y, Johnson T, Walker TS, Wimmer F, Schaldach R, Nelson GC, Willaarts B (2017) Quantitative foresight modeling to inform the

CGIAR research portfolio. Washington, International Food Policy Research Institute (IFPRI)
Sarapura S (2012) Gender analysis for the assessment of innovation processes: the Case of Papa Andina in Peru. Agricultural innovation systems: an investment sourcebook. World Bank, Washington, DC, pp 211–230. 598–602. http://documents1.world-bank.org/curated/pt/140741468336047588/pdf/672070PU-B0EPI0067844B09780821386842.pdf
Sartas M, Schut M, van Schagen B, Thiele G, Proietti C, Leeuwis C (2020) Scaling readiness: concepts, practices, and implementation. CGIAR Research Program on Roots, Tubers and Bananas (RTB). January 2020, 217 p. https://doi.org/10.4160/9789290605324
Scott G (2020) A review of root, tuber and banana crops in developing countries: past, present and future. Int J Food Sci Technol 56:1093–1114. https://doi.org/10.1111/ijfs.14778
Thiele G, Khan A, Heider B, Kroschel J, Harahagazwe D, Andrade M, Bonierbale M, Friedmann M, Gemenet D, Cherinet M, Quiroz R, Faye E, Dangles O (2017) Roots, tubers and bananas: planning and research for climate resilience. Open Agric 2(1):350–361. https://doi.org/10.1515/opag-2017-0039
Thiele G, Dufour D, Vernier P, Mwanga RO, Parker ML, Schulte Geldermann E, Teeken B, Wossen T, Gotor E, Kikulwe E, Tufan H, Sinelle S, Kouakou AM, Friedmann M, Polar V, Hershey C (2020) A review of varietal change in roots, tubers and bananas: consumer preferences and other drivers of adoption and implications for breeding. Int J Food Sci Technol. https://doi.org/10.1111/ijfs.14684
United Nations, Department of Economic and Social Affairs, Population Division (UNDP) (2019) World Population Prospects 2019, Online Edition. Rev. 1. https://population.un.org/wpp/Download/Standard/Population/
Wiebe K, Sulser TB, Dunston S, Rosegrant MW, Fuglie K, Willenbockel D, Nelson GC (2020) Modeling impacts of faster productivity growth to inform the CGIAR initiative on Crops to End Hunger. https://doi.org/10.31235/osf.io/h2g6r
Wossen T, Alene A, Abdoulaye T, Feleke S, Rabbi IY, Manyoung V (2018) Poverty reduction effects of agricultural technology adoption: the case of improved cassava varieties in Nigeria. J Agric Econ 70(2):392–407. https://doi.org/10.1111/1477-9552.12296

2

Collaborative Governance and Management: The RTB Approach

Helen Hambly, Michael Friedmann, Claudio Proietti, Vivian Polar, Sarah Fernandes, and Graham Thiele

Abstract Collaborative programs that facilitate innovation to deliver value at scale require attention to effective program design, management, governance, and leadership. The CGIAR has experimented with different collaborative program design options over its 50-year history, most recently with the CGIAR Research Programs (CRPs) implemented from 2012 to 2021. This chapter examines the structure and processes of the CGIAR Research Program on Roots, Tubers and Bananas (RTB). It unpacks the constituent institutional innovations that underpinned the RTB program, their key design principles, how they evolved over the 10 years of the program, the innovations achieved, and the outcomes to which they contributed. Turbulence and transformations in the CGIAR system influenced the CRPs' emergence, design, and delivery. In this chapter, we discuss the RTB approach to collaborative governance and management as complex institutional innovations operating within this broader, dynamic system. This includes attention to opportunities, limitations, and other contextual factors influencing RTB's work. Institutional innovations include stakeholder consultations and priority setting, a portfolio organized by aggregated innovations, or clusters of activities, articulated flagship projects, incentive funding, a dynamic interactive communication ability, and programmatic embedding of strategic and integrated gender research. RTB's design, governance, and management innovations added value to the combined achievements of the participating centers in science and research for development outcomes, described in the following chapters.

H. Hambly (✉)
University of Guelph, Guelph, ON, Canada
e-mail: hhambly@uoguelph.ca

M. Friedmann · V. Polar · S. Fernandes · G. Thiele
CGIAR Research Program on Roots, Tubers and Bananas (RTB), led by the International Potato Center, Lima, Peru
e-mail: M.Friedmann@cgiar.org; V.Polar@cgiar.org; S.Fernandes@cgiar.org; g.thiele@cgiar.org

C. Proietti
Formerly CGIAR Research Program on Roots, Tubers and Bananas, currently CIRAD, Montpellier, France
e-mail: claudio.proietti@cirad.fr

2.1 Introduction

For 10 years, the CGIAR Research Program on Roots, Tubers and Bananas (RTB) advanced a significant body of scientific knowledge and practice on the clonally propagated staple food crops: banana (principally cooking banana and plantain), cassava, potato, sweetpotato, and yam. Root, tuber, and banana (RT&B) crops have been historically neglected, even until recently (Krishna Bahadur et al. 2018; Almekinders et al. 2019; Scott 2020). RTB's vision was to tap the potential of these crops across the world to improve food security, nutrition, income, and the climate change resilience of smallholders, especially women and youth. RTB brought together the work on RT&B crops across four CGIAR centers (Bioversity, CIAT, CIP, and IITA) and French research partners represented by CIRAD. It sought to strengthen the agri-food system of these crops in partnership with numerous research and scaling organizations. RTB was part of an integrated portfolio of CGIAR Research Programs (CRPs) established in 2012. This chapter explains how RTB developed an integrated set of institutional innovations shaped by a historical context of change in the CGIAR system.

In this chapter, we first briefly address the historical context of the CGIAR and then present the institutional memory of RTB from its emergence in the "New CGIAR" in 2011 to its transition within the "One CGIAR" (2022–2030). RTB had two 5-year phases of program design and management (2012–2016 and 2017–2021). In both phases, the institutional innovation by RTB was critical in developing adaptive capacity and resilience in the context of emerging priorities and resource disruptions. The final sections of the chapter draw lessons of broader relevance for designing international agricultural research for development programs. Highlighted examples of strategic and integrated science outcomes of the RTB program will be covered in more detail in subsequent chapters of this book.

2.2 Context

International agricultural research plays an important role in advancing agricultural science to improve the lives of billions of people in the world, especially the food insecure who face exigencies of poverty and environmental and climate crises. Since the 1970s, coordinating agricultural research for development and its coordination has changed substantially (Dalrymple 2008; Feldman and Biggs 2012; Immonen and Cooksy 2019). Linear models that implicate top-down problem-solving, for example, from scientists to farmers, have shifted toward demand-driven solutions. The inequitable distribution of the benefits of new technologies, for example, seed that is high yielding but requires costly inputs that disadvantage resource-poor farmers, has led to greater recognition of the socio-technical complexity of knowledge plurality and the need for bridging cross-sectoral and disciplinary boundaries (Pigford et al. 2018).

There are advantages to co-produced knowledge and co-innovation with diverse stakeholders in the agricultural system (Pant and Hambly 2009). Approaches to improving innovation have included participatory design process, reflective practice in science management, regional hubs and innovation platforms, learning alliances for scaling, adaptive collaborative management that strengthens local capacity, and partnership projects to foster collective knowledge assets (Horton et al. 2009; Ekboir 2009; Pigford et al. 2018; Berthet et al. 2018). Recently, the COVID-19 pandemic has drawn attention to the role of science and trust in expert knowledge, informing not only research in human diseases but also agri-food systems, climate change, and biodiversity loss (Barrett et al. 2020). The current transition known as One CGIAR (Coffman et al. 2020) is both a response to system changes and a contributor to innovation that can unleash the creativity needed to respond to shocks such as COVID-19 that prioritize donor funding to health.[1]

2.3 CGIAR Collaborative Programs

Created in 1971, the CGIAR is the world's largest group of publicly funded, agricultural research for development organizations. Its creation involved vested geopolitical interests and other powerful forces, creating the context that was to shape the CRPs (Hardin and Collins 1974; Baum and Lejuene 1986; Harwood and Kassam 2003; Ozgediz 2012; McCalla 2014, 2017; Byerlee and Lynam 2020). Early international agricultural research programs, largely funded by the Rockefeller and Ford Foundations, generated positive social returns on investment from partnerships that developed high-yielding crop varieties. From the 1960s onward, growth in and modernization of smallholder agriculture were predicated on new Green Revolution technologies (high-yielding crop varieties, fertilizers, and pesticides), guaranteed commodity prices, subsidized inputs, research, extension, training services, and infrastructure (e.g., irrigation). The evolving mandates of the United Nations Food and Agriculture Organization (FAO), the United Nations Development Programme (UNDP), and the World Bank generated influential global agricultural policy analyses and assessments that led to the CGIAR's creation (Feldman and Biggs 2012). Increased multidisciplinary scientific exchange, civil society debate, and structural adjustment policies of the 1970s–1980s combined to reassess the priorities of the Green Revolution, which was argued to have typically benefited wealthier, larger-scale, and male farmers. This led to greater attention to equity, including gender and interventions directed toward less well-off smallholder farmers (Glaeser 1987; Feldman and Biggs 2012).

By the early 1990s, the CGIAR's first four decades were strained by the complexity of taking on issues from the global agenda set out in the World Commission on Sustainable Development (WCSD). This included adding new CGIAR centers,

[1] See also https://www.cultivation.hps.cam.ac.uk/CGIAR-histories.

which grew from 4 in 1971 to 12 in 1979. Limited funding led to competitive stress and coordination challenges. The CGIAR centers had, up to the 2009 reform era, two forms of funding: unrestricted (referred to as core funding provided jointly by the World Bank and other donors) and restricted funds (bilateral funding from specific donors for projects). The former supported centers' management costs, allowing operational flexibility when new cost items emerged. The latter covered donor-specified costs in projects and centers. Unrestricted/core funding declined relative to restricted and special project funding, from upwards of 80% in the first decades of the CGIAR (Operations Evaluation Department 2004) to about 30% of the system funds by its fourth decade (Ozgediz 2012). Consequently, some centers faced financial difficulties, as they competed for project funds and had little appetite for the unfunded transaction costs of system-wide initiatives (McCalla 2014, 2017). Added to this was a degree of protectionist behavior in the system that resisted further merging or a reduction in the number of centers.[2]

New priorities reflected in the WCSD led the CGIAR to give increasing attention to the environment, biodiversity, and sustainability. Against this backdrop, reform sought a strategy of unification respecting centers' autonomy with large, coordinated multicenter research programs to tackle these global issues (Fig. 2.1). This led to the creation from 1993 of system-wide and eco-regional and CGIAR initiatives involving attention to greater interdisciplinary farming systems research, farmer participatory research, location-specific research, and new centers focused on natural resource management and biodiversity (Greenland 1997). By 2002, a mechanism to invite collaborative proposals on large, system cross-cutting research programs emerged. These 5 Challenge Programs (CPs) aimed to improve CGIAR business processes while shifting the consultative group of centers toward a

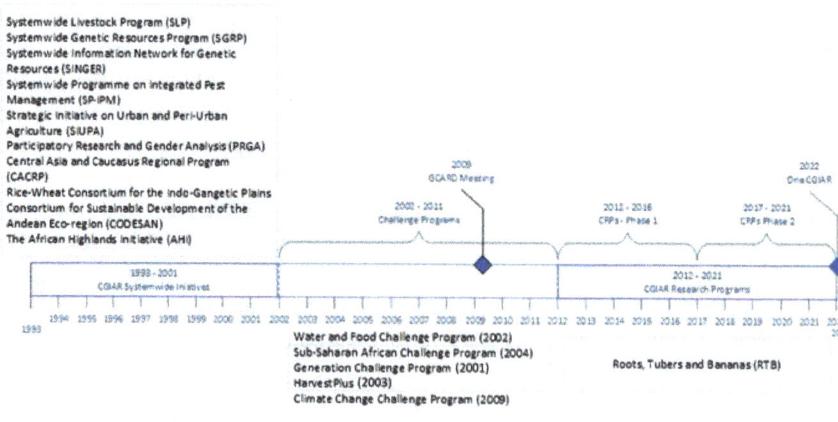

Fig. 2.1 Timeline for CGIAR collaborative programs with examples

[2] ILCA and ILRAD merged to become ILRI. INIBAP (banana and plantain) joined IPGRI (later Bioversity). ISNAR closed with some activities folded into IFPRI.

"consortium" model of setting system goals and regional priorities and enabling co-funding to reduce growing dependency on restricted funding (TAC/Science Council 2003).[3] One key element to engage partners in each CP was independent priority setting and management oversight bodies independent from the centers (Woolley et al. 2011).

This formative period from 1981 to 2010 prepared the ground for what would be the future cross-cutting CRPs (Fig. 2.1). Nevertheless, despite the desire for integration, reflected in new system-wide initiatives, the CGIAR center-based model of collaborative international agricultural research for development not only survived but also grew, with scientific work by the centers having a high impact.[4]

Seeking further changes toward more coherent responses to the United Nations' Sustainable Development Goals (SDGs) reduced fragmentation of the centers, and to avoid duplication of efforts among them and to optimize available funding, CGIAR funders and centers came together for a major reform effort in 2009. In 2010, delegates of the Global Conference on Agricultural Research for Development (GCARD), in Montpellier, France, helped shape a *Road Map to Transform the Agricultural Research for Development (AR4D) System for Greater Global Impacts* (GCARD 2011). The conference finalized a new Strategic Research Framework (SRF) for the CGIAR integrating the work of all the centers in a results-oriented research for development system with an integrated programmatic structure (Consortium Board 2011).

2.4 Design and Evolution of CRPs

The proposed building blocks of the SRF were a set of (initially) 15 interdependent CGIAR Research Programs (CRPs). The CRPs were envisaged as multicenter, interdisciplinary, and collaborative results-oriented programs whose impact was expected to be greater than the sum of their parts, because of the gains from synergies and system-wide cooperation (Consortium Board 2011). Donors responded favorably to the SRF, creating pooled funding through the mechanism of the "CGIAR Trust Fund," designed primarily to finance centers' collaboration and AR4D synergies (Renkow and Byerlee 2010) (Table 2.1).

[3] The World Bank initially hosted the CGIAR Secretariat, and the FAO hosted the Technical Advisory Council (TAC) in Rome, which held science impact, shared research service platforms, and other accountability functions. Under the consortium, the System Office was a virtual entity, and later the Consortium Office was based in Montpellier. TAC was renamed the Science Council retaining its advisory and evaluative functions.

[4] The impact of the CGIAR is summarized in two reports: at its 31-year milestone (Operations Evaluation Department 2004) and at 40 years (Ozgediz 2012). For example, the latter report stated that every dollar invested in CGIAR research meant $9 worth of additional food was produced in developing countries.

The Fund was intended to pledge stable, long-term financing for collaboration and integration of activities across CGIAR centers. W2 gave funders the discretion to invest in particular CRPs or platforms. However, an internal stabilization mechanism allowed for compensating allocations through W1 when there was a reduction in W2 allocation for any CRP. This created more stable funding from year to year but somewhat frustrated the funders' intentions to support those research areas they deemed of the highest relevance.

Initially, with the enthusiasm for the new reform, funder investments in the CGIAR rose markedly, climbing above $1 billion in 2014. However, as difficulties appeared, funding fell to $800 million in 2019. With this promising start, pooled funding (W1&2) reached 35% of total CGIAR funding but then fell sharply through 2015–2017 to around 20% of the total (Fig. 2.2). The CRPs would face challenges responding to funding-related turbulence and the late confirmation of funding within each financial year.

Even midway through SRF Phase 1 in 2015, there was a growing feeling among donors that the portfolio was too complex, inhibiting the desired collaboration. A decision was taken to close the system's CRPs and reduce from the original 15 to 11 CRPs (CGIAR 2015a, 2016). Building on strong prior collaboration with RTB, much of the CRP on Integrated Systems for the Humid Tropics (CRP-Humidtropics 2016) that was led by IITA, including the work in innovation platforms and their budgets, was integrated into RTB. This incorporation made RTB the largest Phase 2 CRP in the system portfolio by 2017.

The CGIAR portfolio was redesigned in preparation for a Phase 2 beginning in 2017. This standardized the programmatic structures for each CRP, based on flagship projects (FPs), and established a clear framework for cross-CRP integration. The design process had two stages: (a) pre-proposals and (b) full proposals, each evaluated by the Independent Science and Partnership Council (ISPC). This significantly strengthened CRP and system-level consistency across the CGIAR portfolio compared to Phase 1, with common design principles and carefully crafted theories of change to which the FPs contributed (ISPC 2012). The new portfolio was

Table 2.1 CGIAR funding windows, 2012–2021

Funding window	Purpose
Window 1 (W1)	Portfolio investments: funding allocated to the entire CGIAR portfolio of approved system-wide investments, prioritized and allocated by funders collectively through the System Council – supporting CGIAR as a whole
Window 2 (W2)	Program investments: funding allocated by funders individually to any component (e.g., CRP, Platform) of the system-wide portfolio as prioritized, defined and approved by the funders collectively through the System Council
Window 3 (W3)	Project investments: funding allocated by funders individually to projects that are defined by the funders themselves (with partners) and that are aligned with system-wide investments

Source: CGIAR website

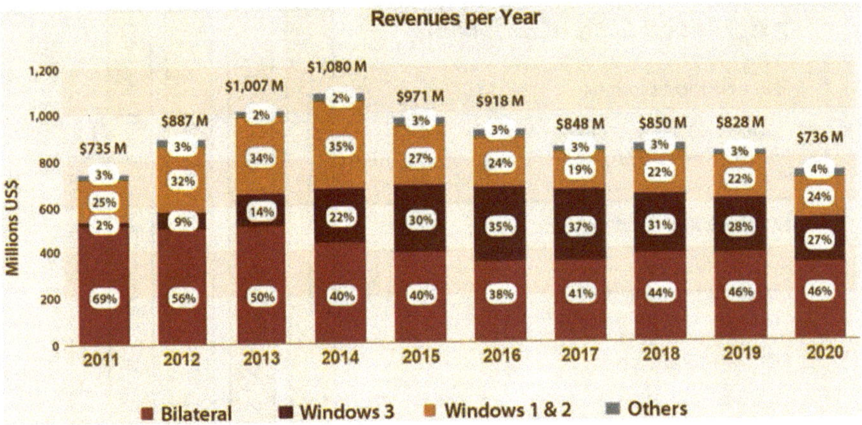

Fig. 2.2 CGIAR revenues per year

comprised of seven agri-food system programs and four interlocking globally integrating programs (Fig. 2.3).

2.5 RTB Program

The planning and evolution of the RTB program is part of the long history of CGIAR reform described above, comprising three phases of activity: a preparatory design phase (2009–2011), Phase 1 (2012–2016), and Phase 2 (2017–2021), summarized in Fig. 2.4.

During its design and Phase 1, RTB was conceived as one of the commodity CRPs, which aimed at modernizing crop breeding programs, creating synergies based on the commonalities of clonal crops and linking breeding to seed systems; these domains made up about 40% of total investment. In the design phase, a group of scientists from four CGIAR centers,[5] CIP (as the lead center), with Bioversity, CIAT, and IITA, met up to develop a scope of work and identify research components of the participating CGIAR centers (CRP-RTB 2011).

To design its program impact pathway and ensure consistency with the system-level outcomes, RTB sought to link relevant center activities at multiple levels, from farm households to wider production systems, and across national, international, and regional levels with diverse public and private sector stakeholders. A central part of the impact pathway focused on programmatic integration across CGIAR

[5] Following a scoping analysis, led by the RTB Steering Committee, other strategic science partners and CIRAD (a non-CGIAR center representing the French research partners) were subsequently invited to join the consortium.

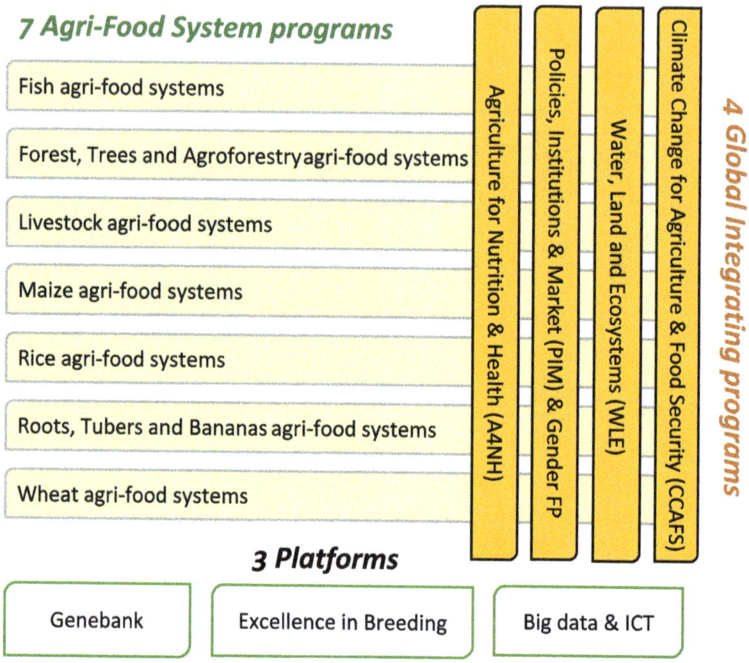

Fig. 2.3 Comprehensive and integrated CGIAR portfolio for Phase 2

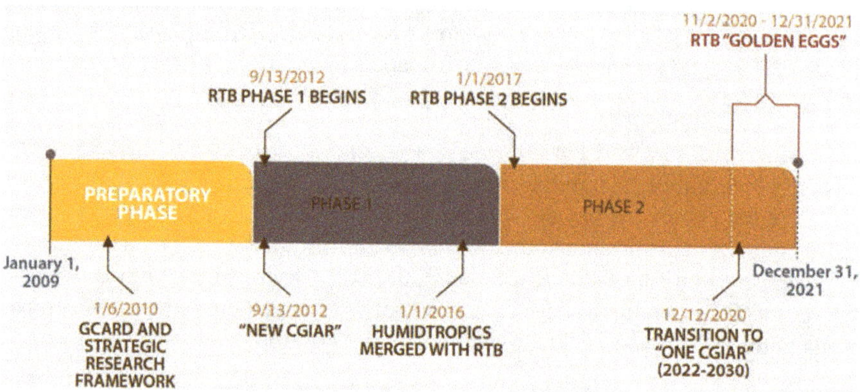

Fig. 2.4 RTB timeline (2009–2021)

core skills, adding value and capturing synergies across the five prioritized[6] crops (banana, cassava, potato, sweetpotato, and yam) and across the five research centers.

[6] RTB also included the minor root and tuber crops such as taro and several Andean crops. However, these were not prioritized for pooled funding despite their importance in some localities.

During implementation in Phase 1, new AR4D partnerships were created through RTB's strategic use of the W1/W2 funds. One outstanding example that began in the first phase was for work on metabolomics across all crops with Royal Holloway, University of London (Price et al. 2020). RTB also linked with other CRPs, in particular to Humidtropics, which was noted in its 2016 annual report:

> Tremendous success was made towards the incorporation and mainstreaming of components of Humidtropics research within some AFS-CRPs. The biggest success in this regard was in relation to partnership with CRP Roots, Tubers and Bananas (RTB) Agri Food Systems. RTB was open to partnership with Humidtropics and took on a number of research activities including Innovation Platforms and place-based systems research operations in a number of locations. Humidtropics contributed in the mainstreaming of systems approaches and the development of a Livelihoods Flagship within RTB.[7]

Against this backdrop of continuous reform, RTB developed as a complex institutional innovation that made purposeful change in governance and program management, drawing on "soft skills" (Woodhill 2010) such as communication, trust building, networking, and leadership with clear goal orientation. This enabled enhanced collaboration by cultivating and pursuing collective action against shared goals, which is highly challenging if institutions and organizations work alone (Roberts and Bradley 1991; Arena et al. 2017). Early on, RTB management recognized that collaborative practices were key to driving innovation and responding strategically to shifts in the external environment. The main institutional innovations that facilitated collaborative practices, described in this section, are the following:

1. Collective action in management, leadership, and associative governance
2. Stakeholder consultation and participatory design
3. Priority setting to guide investments and build/adapt the portfolio
4. Portfolio organized by aggregated innovations
5. Programmatic embedding of strategic and integrated gender research
6. Internal funding mechanisms and incentives
7. Dynamic interactive communication capability to build a shared vision and stakeholders' support
8. Purposive engagement of national partners

2.5.1 Collective Action in Management, Leadership, and Associative Governance

Early on, during the design phase, the centers involved in RTB opted for an associative style of governance. In terms of a "hard contract," CIP is the lead center, and its Board of Trustees holds fiduciary responsibility for RTB. Yet with an equally important "soft contract" approach, through the management committee, participating centers in RTB were given an equal weight in taking decisions with transparent access to information, particularly on finances. The different management and governance levels in RTB recognized that mechanisms were needed to support collective action, which could not be achieved by any of the centers acting separately (Horton et al. 2009). An associative style of governance was both a reason for and a result of collective action in RTB. Each participating center in RTB was an important contributor to the overall impact of clonally propagated crops, and bananas and cassava had mandates shared by two centers.

In Phase 1, RTB governance involved a steering committee comprised only of center DGs and a science advisory committee known as the PAC (Program Advisory Committee) whose membership included subject and regional specialists from Non-Governmental Organizations (NGOs), universities, and research institutes around the world (Fig. 2.5). The PAC played a role as part of the system of checks and balances by looking out for collective interests above those of each of the centers.

Operationalizing collaboration was facilitated by the full-time RTB Program Director (PD) supported by the Project Management Unit (PMU) in close communication with the RTB Management Committee composed of apex research managers in the participating CGIAR centers (usually Deputy Director General of Research). The PD led the Program Management Unit (PMU) that included five to seven full-time support staff responsible for grants, finance, communications, planning and reporting, and gender research, with a science officer included from 2015.

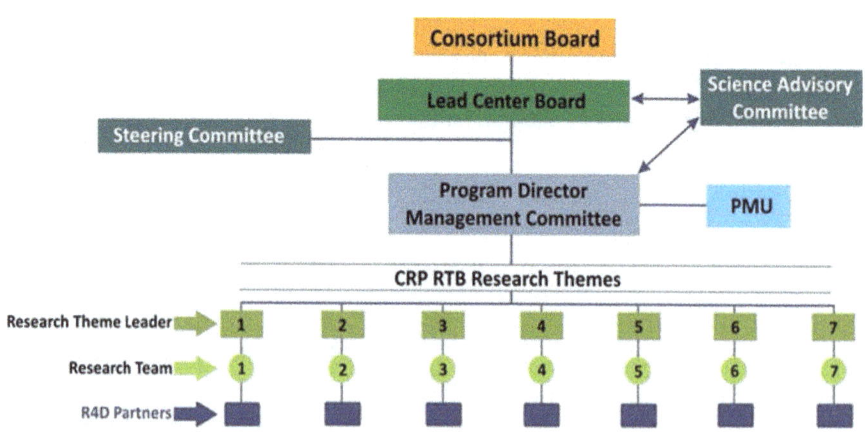

Fig. 2.5 RTB organogram Phase 1

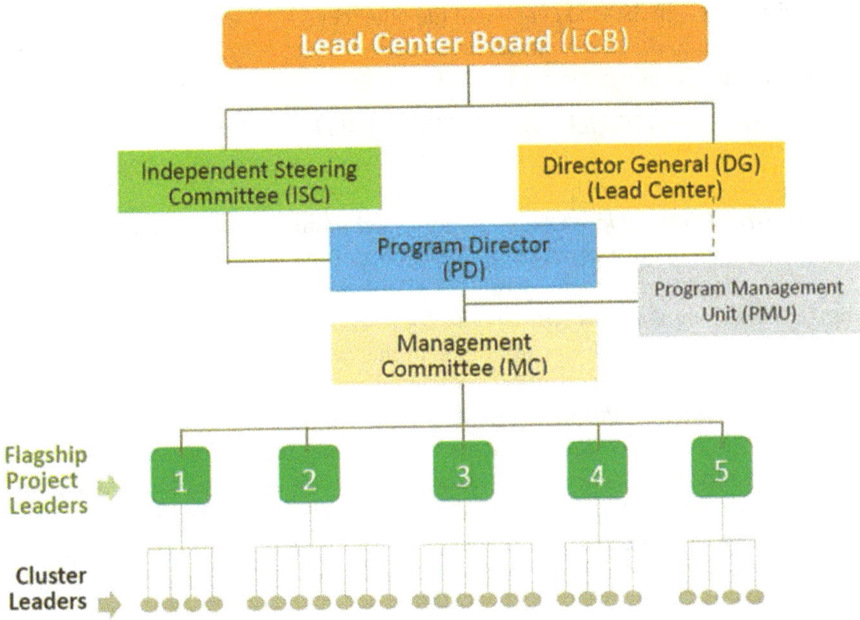

Fig. 2.6 RTB organogram for Phase 2 showing governance and oversight functions

The PD, guided by the Management Committee, oversaw FP leaders and the cluster leaders with a programmatic reporting line (Figs. 2.5 and 2.6). Reflecting the soft contract approach, FP and cluster leaders were drawn in a balanced way from all five participating centers.

During Phase 1, the RTB collaborative governance structure evolved further, following system guidelines for all CRPs, into the Independent Steering Committee (ISC, formerly the Program Advisory Committee) where the center DGs are represented, but most independent members are selected according to needed competencies for program oversight while ensuring some level of regional and gender representation. The ISC Terms of Reference specified that it had advisory input to the RTB to ensure oversight of the strategic alignment of the RTB with the SRF at crucial periods such as the design of CRP-level program proposals, draft annual Plan of Work and Budget (PoWB), and Annual Reports. The Chair of ISC reported annually to the CIP Board of Trustees, which had fiduciary responsibility for RTB. The Program Committee of the CIP Board received regular reports on program progress and approved the RTB annual PoWB. RTB developed a comprehensive and coordinated governance function with collaborative practices such as collective review of the PoWB and the review of FPs' progress and plans as a central feature of ISC annual meetings.

RTB's associative style of governance and collaborative management was further reinforced following a recommendation from the Independent Evaluation Arrangement (IEA) review (2016) to create an "alliance compact" as a trust-built, soft contract among the participating centers to bolster the legal agreements established between the lead center and the other centers. The ISC played a key role in translating the alliance compact into a specific set of partnership statements, signed by the DGs of the centers in 2017, as follows:

(a) *RTB Partnership Collaboration* – RTB is a shared asset that should be jointly promoted and nurtured in a collaborative way to support collective action and add value to members in the long term.
(b) *Inclusive Partnership* – Openness, trust, and mutual respect and learning lie at the core of effective partnerships in support of the RTB goals, recognizing the different and complementary roles of all members.
(c) *Strengthening Business Partnerships* for RTB for cross-cutting, multi-crop, collective action research for development that would not be possible by each of the members acting separately.
(d) *Donor Relations* – Promoting joint stewardship to maintain the engagement of the existing set of Window 2 donors and sharing responsibilities for resource mobilization intelligence for potential new Window 2 funding for RTB. Centers agree to pursue a policy of minimal reciprocal overheads for pass-throughs that relate to transfers across the members. For Window 3 or bilateral funding connected to the RTB Program, members agree to map the funding and results into the RTB Program so that the program as a whole could benefit.
(e) *Talent Management* that flows across centers for RTB management positions such as FP and cluster leaders is a key part of the overall compliance mechanism.
(f) *Communications* – Ensure that communication/public relations activities accurately reflected collaborative efforts and the contribution of each member.

In summary, this associative governance style supported major transitions within the first and second phases of RTB, including after 2016, the incorporation of parts of the Humidtropics CRP. The soft contract of the "alliance compact" recommended by IEA was acted upon, and a collective approach to management and governance was adopted.

2.5.2 *Stakeholder Consultation and Participatory Design*

Stakeholder consultation was an institutional innovation supporting the design process, particularly in the preparatory phase of RTB. There was a short timeframe between the start of proposal writing (late June 2010) and the deadline for submitting the proposal (the first week of September 2010); 255 stakeholders, about half from developing-country national agricultural research systems (NARS) and universities, were consulted using surveys. This resulted in an initial RTB structure with seven disciplinary themes, each with a mix of existing, expanded, and new product lines as well as cross-cutting activities. To create its initial structure, more than 25 researchers from Bioversity, CIAT, CIP, and IITA participated in a 3-day workshop, held at CIP's headquarters in 2010, to define and organize a strategy for developing the proposal for a CRP. Writing teams were formed across topics to

Table 2.2 RTB stakeholders' aggregated score of importance assigned to seven themes in the proposal

Theme	Regional survey	Global survey
Theme 2: Accelerating the development, delivery, and adoption of varieties with stable yields, stress resistance, and high nutritional value	4.60	4.55
Theme 6: Enhancing postharvest technologies and adding value in markets	4.58	4.22
Theme 4: Promoting sustainable systems for clean planting material for farmers	4.51	4.38
Theme 1: Conserving and accessing genetic resources	4.42	3.81
Theme 3: Managing priority pests and diseases and beneficial microbial communities	4.29	4.24
Theme 7: Enhancing impact through partnerships	4.33	4.00
Theme 5: Developing tools for more productive, ecologically robust crops	4.12	4.05

Note: Themes were scored on a 0–5 scale (0 meant "not important" and 5 was "very important")

encourage cross-center collaboration, and writing responsibilities were assigned. A proposal development schedule was developed and agreed upon, as was a protocol for writing, editing, and managing the draft sections. The inter-center workshop proposed seven core themes for RTB (Table 2.2).

The seven CRP-RTB themes were ratified by the stakeholders (Table 2.2). The stakeholder surveys during the design phase confirmed previous findings that RTB crops are generally absent from government rural development strategies (Woolley et al. 2011). Following the design workshop and stakeholder consultation, RTB was organized by these seven themes. Each theme was ordered by crop. Each theme leader was drawn from a different center, and each theme included cross-cutting work in gender, biophysical research, and processes and partnerships for scaling innovations.

By 2014, with the implementation of results-based management in the CGIAR, and more consideration of its theory of change, the PMU realized that the first thematic structure (labelled RTB 1.0) had limitations for creating a compelling and viable theory of change and for organizing science teams. RTB 1.0 neatly arranged work by different themes but lacked a process to bridge the different themes, e.g., connecting varietal development with seed system development for each crop.[8] Working groups in the 2014 RTB annual meeting in Kampala considered options for a new program structure (referred to as RTB 2.0). Before the start of Phase 2, which began in 2017, the RTB portfolio had been reorganized into five FPs and clusters of activities (CoA) that encompassed five crops, four CGIAR centers, and by this point a non-CGIAR center, CIRAD (CGIAR-IEA 2015a, b, 2016). The

[7]The analogy was that RTB 1.0 was like a closet with separate groups of clothes on hangers, but no way to combine them into complete and matching outfits.

transition toward the RTB 2.0 structure, which replaced seven themes with five FPs, was gradually completed in 2016.

The proposal for Phase 2 (RTB 2016a) including a scope of work and a theory of change for each of the five FPs was prepared by design teams drawn from RTB scientists from each FP (see Sect. 2.5.5). Key staff from the Humidtropics CRP joined RTB on these writing teams, particularly around scaling innovation (FP5), which included researchers from Wageningen University & Research (WUR). Additional teams worked on cross-cutting topics of gender-responsive research, capacity development, and partnerships (Fig. 2.7). This proposal combining flagships and cross-cutting topics was reviewed and approved by the ISPC in a two-stage submission process. RTB received one of the highest ratings of any CRP by the ISPC (2016).

In summary, RTB 1.0 became RTB 2.0 by supporting the interaction between innovations, which is essential for wider adoption and scaling (see Sect. 2.5.4). At the same time, there was extensive CGIAR consultation during 2014–2016 that restructured all CRPs with shared design and management principles around FPs and clusters for Phase 2.

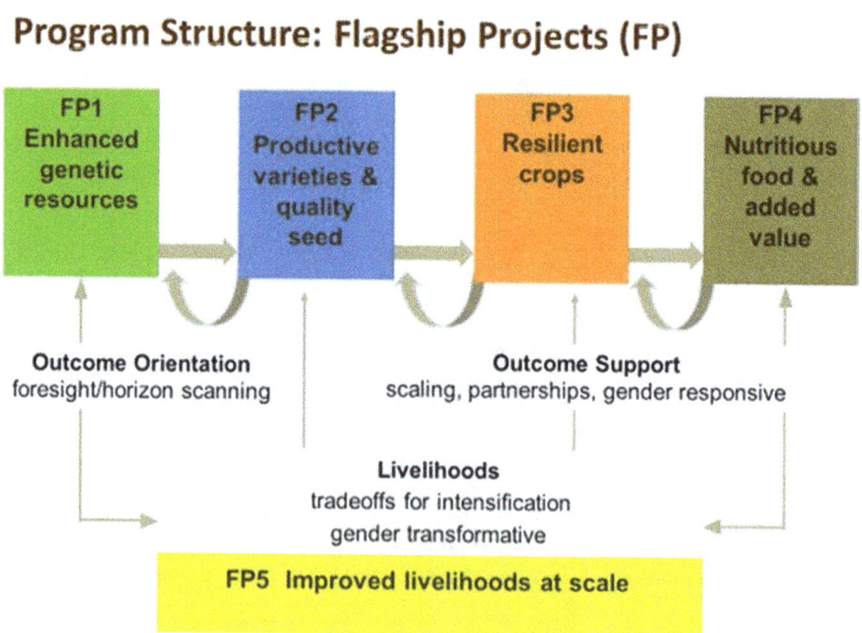

Fig. 2.7 Program structure organized by FPs, 2016–2021 – RTB 2.0

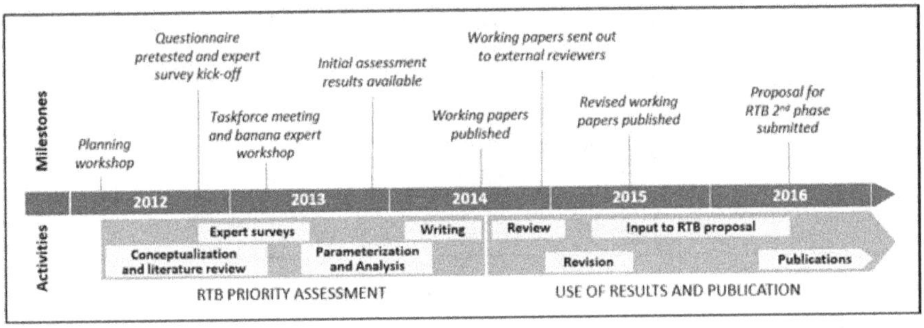

Fig. 2.8 Timeline and key activities of RTB priority assessment. (Source: Pemsl et al. 2022)

2.5.3 Priority Setting to Guide Investments and Build the Portfolio

Reflecting on the stakeholder consultations in 2010, RTB responded to a direct request from donors and the ISPC to conduct a rigorous priority assessment as a precondition for funding. The steps of the priority assessment method ran in parallel with the participatory design/redesign of the portfolio and the comprehensive framework, including the impact pathways described above (Fig. 2.8).

Priority setting in RTB covered five crops (cassava, banana, potato, sweetpotato, and yams) using a harmonized method that drew on the approach developed earlier at CIP (Fuglie and Thiele 2009). The assessment included four key steps: (1) elicitation of major production constraints and research opportunities through global and regional expert surveys; (2) identification of priority research interventions by crop; (3) ex ante estimation of costs and benefits for two adoption scenarios using partial equilibrium economic surplus models with poverty impact simulations; and (4) publication and use of findings. For the first step, to identify key constraints and research opportunities, 1709 experts were consulted, from a wide variety of disciplinary backgrounds, mostly scientists from national agricultural research organizations.

Specific research options were selected based on overall and regional scores from the expert survey (step 2) in consultation with experts and center scientists, considering the scope of RTB's research activities to ensure a good match of options with the program portfolio.

The results (Table 2.3) showed large benefits for all potential research investments and provided useful outcome and impact indicators (adoption area, number of beneficiaries, net present value, internal rate of return, and poverty reduction). In Table 2.4, results of computed performance indicators suggest two key findings: (1) expected adoption areas are large, ranging from several hundred thousand to almost four million hectares, which translates into high numbers of beneficiaries (estimated to be as high as 36 million persons), and (2) all options have large positive net present values. There was a high level of congruence between the research options of highest priority in the assessment and the crop-specific clusters in RTB's

Table 2.3 Results of RTB priority assessment – adoption, beneficiaries, economic benefits, and poverty impacts (lower adoption scenario)

Top ranked research options by crop	Adoption area	Number of beneficiaries		Net present value (NPV)	Internal rate of return (IRR)	Poverty reduction
	(million ha)	(million HH)	(million persons)	(US$ million)	(%)	(million persons)
Banana						
Banana Xanthomonas wilt (BXW) management cultural practices	0.64	3.22	15.67	1982	72	1.61.
Recovery from banana bunchy top virus (BBTV)	0.40	2.02	9.67	1337	61	0.64
Resistant plantain (RELEASE)	0.45	1.70	7.57	1111	64	0.25
Cassava						
High-quality planting material production and distribution systems for improved varieties	3.38	6.73	33.08	7585	416	2.10
Sustainable crop and soil fertility management	3.27	6.43	31.72	8284	210	2.66
High yielding, drought-tolerant varieties and increased water-use efficiency	3.99	7.89	36.49	3025	61	2.00
Potato						
Late blight resistance	0.77	6.73	33.08	7585	416	2.10
Virus-resistant varieties	0.36	6.43	31.72	8284	210	2.66
Bacterial wilt-resistant varieties	0.64	1.72	7.85	253	29	0.20
Sweetpotato						
Orange-flesh sweetpotato (OFSP)[a]	0.67	3.00	14.60	563	35	0.48
Weevil-resistant varieties	0.72	2.94	14.11	363	41	0.36
Sweetpotato virus disease (SPVD)-resistant varieties	0.48	1.96	9.41	673	116	0.34
Yam						
Clean planning materials and agronomic practices	0.68	2.39	17.72	570	37	0.18

(continued)

Table 2.3 (continued)

Top ranked research options by crop	Adoption area (million ha)	Number of beneficiaries (million HH)	(million persons)	Net present value (NPV) (US$ million)	Internal rate of return (IRR) (%)	Poverty reduction (million persons)
Improved varieties with complementary ICM	0.43	1.58	11.74	3026	60	0.66
Yam pest and disease management options	0.43	1.60	11.85	412	43	0.10

Source: Pemsl et al. (2022)
[a]Including health benefits from the adoption of OFSP (DALY method) substantially increases benefits: NPV: US$1298 million, IRR: 51% (lower adoption scenario)

Table 2.4 RTB portfolio organized by clusters or "innovation packages"

Discovery	Delivery		
FP1: Enhanced genetic resources	FP2: Productive varieties and quality seed	FP3: Resilient crops	FP4: Nutritious food and added value
DI1.1 Breeding CoP **DI1.2** Next generation breeding **DI1.3** Game-changing traits **DI1.4** Genetic diversity	**CC2.1** Access to quality seeds/varieties **BA2.2** User-preferred banana cultivars/hybrids **CA2.3** Added value cassava varieties **PO2.4** Seed potato for Africa **PO2.5** Potato varieties for Asia **SW2.6** User-preferred sweetpotato varieties	**CC3.1** Pest/disease management **CC3.2** Crop production systems **BA3.3** Banana fungal and bacterial wilts (Foc/BXW) **BA3.4** Banana viral diseases (BBTD) **CA3.5** Cassava biological constraints, Asia/Americas **CA3.6** Cassava biological threats, Africa	**CC4.1** Post-harvest innovation **CA4.2** Cassava processing **CA4.3** Biofortified cassava **SW4.4** Nutritious sweetpotato
FP5: Improved livelihoods at scale			
CC5.1 Foresight and impact assessment **CC5.2** Sustainable intensification and diversification for improvement resilience, nutrition, and income **CC5.3** Gender equitable development and youth employment **CC5.4** Scaling RTB agri-food system innovations			

programmatic structure for FPs 2, 3, and 4 where most delivery research is located (Pemsl et al. 2022). Hence, the exercise gave coherence to the selection of crop clusters for Phase 2, discussed below. The findings informed the development of RTB's research portfolio and were critical for facilitating continued program funding.

In summary, RTB responded to a direct request from donors and the ISPC to conduct a rigorous priority assessment as a precondition for funding. The entire exercise brought social scientists, breeders, agronomists, and other disciplines from different centers together around a shared task and method to provide consistent metrics across crops. Establishing such a community of practitioners was an

institutional innovation, which provided a strong, collaborative base and set an example for cross-center cooperation.

2.5.4 *RTB 2.0: Portfolio Organized by Aggregated Innovations with Linked Impact Pathways*

After 2016, RTB 2.0 was used as a comprehensive framework for planning, implementing, reporting, and learning. FPs were organized into 25 clusters of activities comprising the full RTB CRP portfolio. The three types of clusters were the following:

- Discovery clusters that included the upstream work, feeding into the crop-specific clusters.
- Crop-specific delivery clusters generating direct impact
- Cross-cutting clusters that synthesized, linked, and supported work across the crop-specific clusters

The cross-cutting clusters are a key institutional innovation as they contribute to establishing communities of practice across different crops and centers. Few other CRPs had such clusters (Jill Lenné, personal communication).

Each cluster was designed with a lead or core innovation (lead product) and an array of complementary innovations (linked products). The clusters sought to include multidisciplinary expertise from the RTB themes. Each cluster was designed by a science team that was responsible for developing the innovations and preparing a business case for the cluster that demonstrated the value added (Table 2.4).

The lead innovation[9] is the centerpiece of a work package that also consists of complementary, linked, or enabling products and includes a theory of change with quantified indicators (RTB 2013b). A lead innovation was defined as:

1. A significant measurable and time-bound product (including knowledge, technology, and organizational and institutional models) that results from a research activity or set of related activities attributable to RTB
2. Used by a well-defined group of next users who may be researchers or development actors, with strong evidence of demand-pull from these users
3. Near market-ready set of ideas, technologies, or science products that generate excitement among researchers and other users
4. With potential for contributing to large-scale impact

RTB 2.0 recognizes that a single innovation cannot be used at scale on its own as it requires complementary innovations for broader use (Sartas et al. 2020). This is illustrated by the seed potato for Africa cluster (Fig. 2.9). "Business models for

---

[8]The lead innovation was referred to as a flagship product or flagship products. We use the term "lead innovation" here so as not to be confused with FPs.

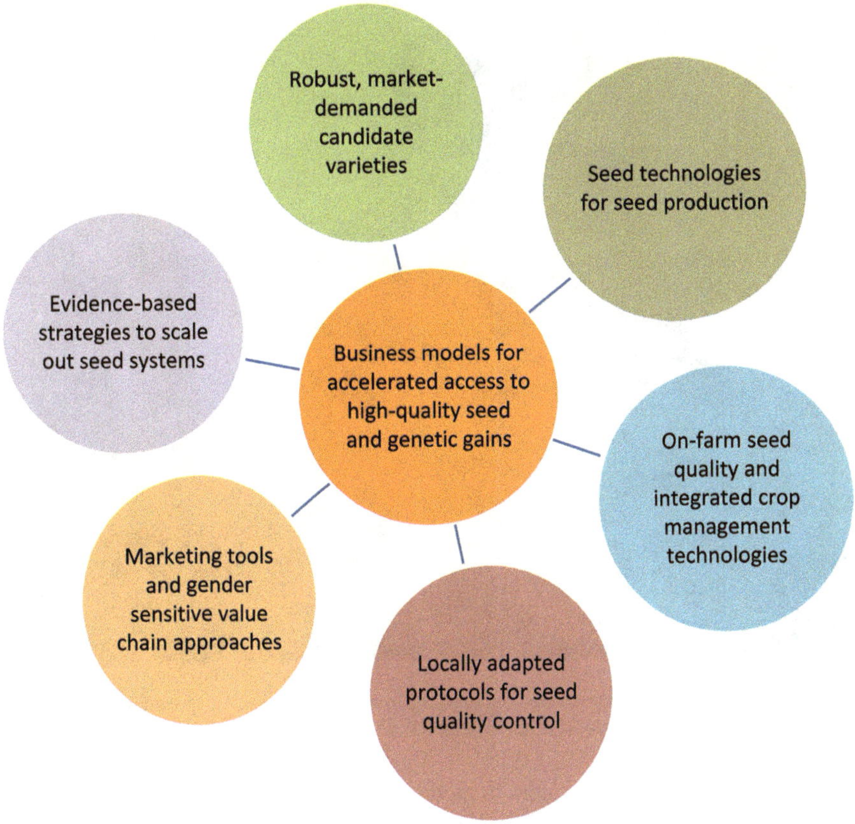

Fig. 2.9 Lead innovation and complementary innovations for the cluster of seed potato for Africa

fostering farmers' access to seed" was the lead innovation that could facilitate farmer access to market-demanded varieties and would require new seed production and quality management technologies, gender-sensitive approaches for value chain development, and scaling strategies. Thus, by organizing each cluster as aggregated innovations, connections could be made to the theory of change developed for the FP, potentially bringing together the bilateral projects mapped under the cluster of activities.

RTB 2.0 paid attention to impact pathways linked to the Intermediate Development Outcomes (IDOs) as set out in the CGIAR Strategic Research Framework (SRF) (CGIAR 2015b). Impact pathways were not drafted by one individual at a desk but were designed with the FP and cluster teams and articulated to the SRF IDOs. Where possible, RTB consulted with stakeholders, using participatory and reflective tools like pathway visualization (Fig. 2.10) to develop impact pathways (Fig. 2.11). During 2014 and 2015, as part of the piloting of CGIAR results-based management, RTB held workshops with a broad group of stakeholders for a detailed validation and co-construction of four cluster impact pathways. Some clusters identified

Fig. 2.10 Stakeholders co-constructed the impact pathway for potato quality seed in Africa during the workshop in Kenya in 2014. (Photo: G. Thiele)

quantified performance indicators to guide results-based management based on the impact pathway (Fig. 2.12). RTB did not, however, have sufficient resources to implement the monitoring and evaluation (M&E) system as initially planned with all of the cluster teams, considering the intensive convening and facilitation required.

RTB's 2.0 comprehensive framework also supported institutional innovations for monitoring and reporting. A major improvement introduced in Phase 2 was the use of the Monitoring, Evaluation, and Learning (MEL) software platform co-developed among several CGIAR centers. MEL allowed systematic tracking of progress and aggregation of results across the RTB and other CRP portfolios. Figure 2.13 illustrates the indicators and case studies included at the output, outcome, and impact level. MEL used a generic description of the program's scope of control (output), scope of influence (research outcomes with next users), and scope of interest (development outcomes with end users) developed by the CGIAR System Office. MEL enabled reporting through a CGIAR-wide dashboard.[10]

Initially, MEL had limitations for comprehensive results-based management. First, it did not show the progress of innovations moving from one state of readiness to the next in successive years. Second, it did not adequately capture the aggregated innovations within a cluster. Third, there was an issue of synchronicity, as the

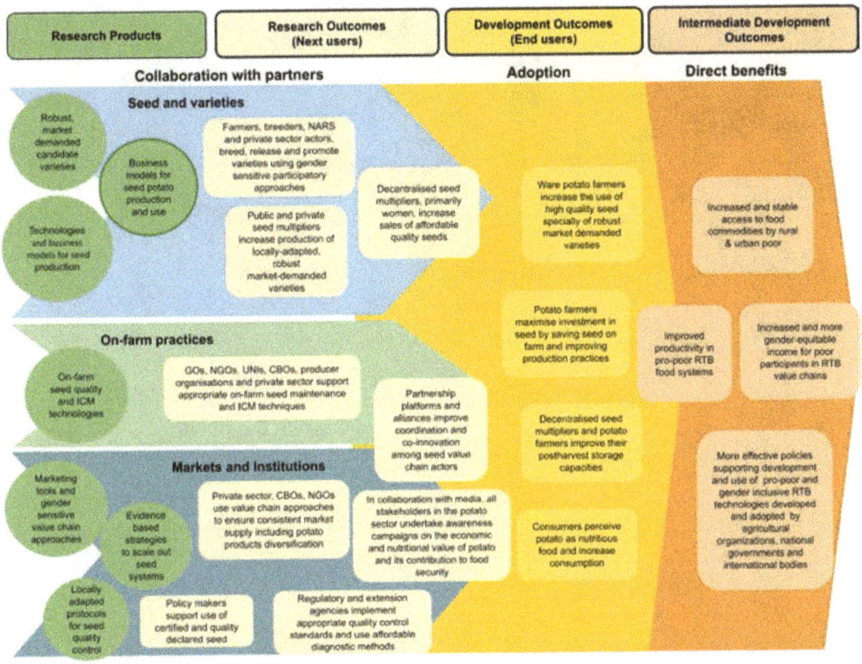

Fig. 2.11 Co-constructed impact pathway for seed potato in SSA

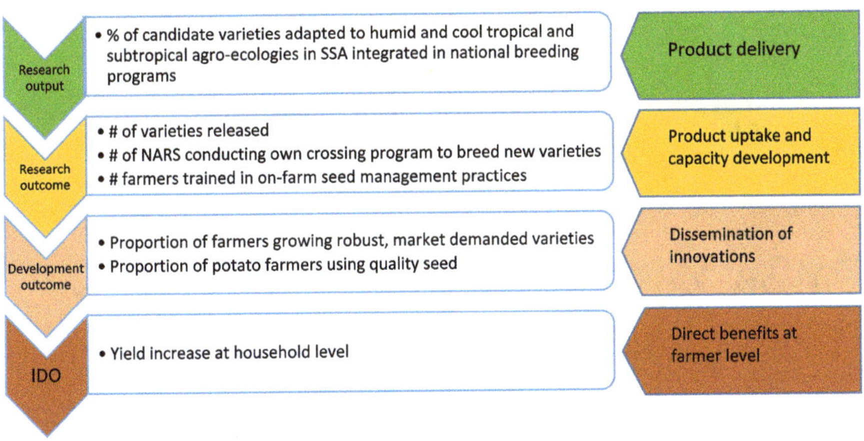

Fig. 2.12 Variables initially proposed for tracking along the impact pathway for seed potato in SSA

outcome cases and the adoption and impact studies reported in any year related to innovations developed earlier, sometimes much earlier. MEL was improved, and from 2018, in coordination with the CGIAR System Office, the readiness of each innovation was assessed, and the progress of the innovations was tracked as they moved from one level of readiness to the next. This permitted a visualization of

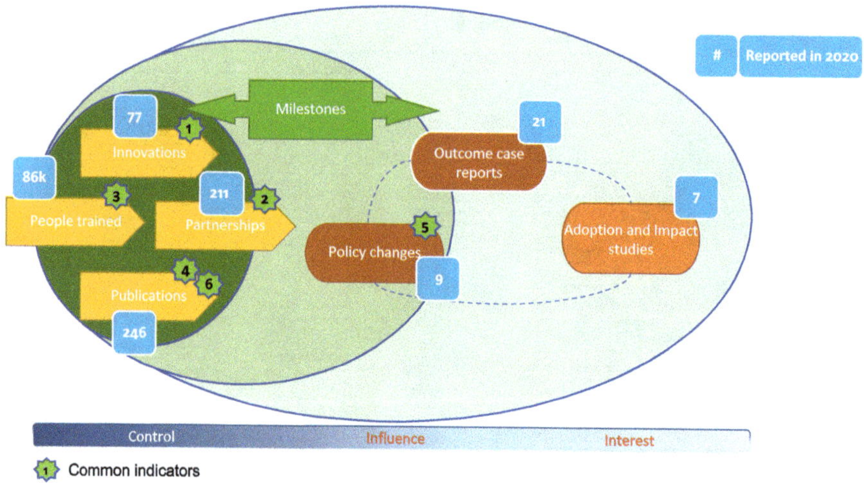

Fig. 2.13 Systematic reporting of deliverables, research outcomes, and development outcomes with MEL in 2020

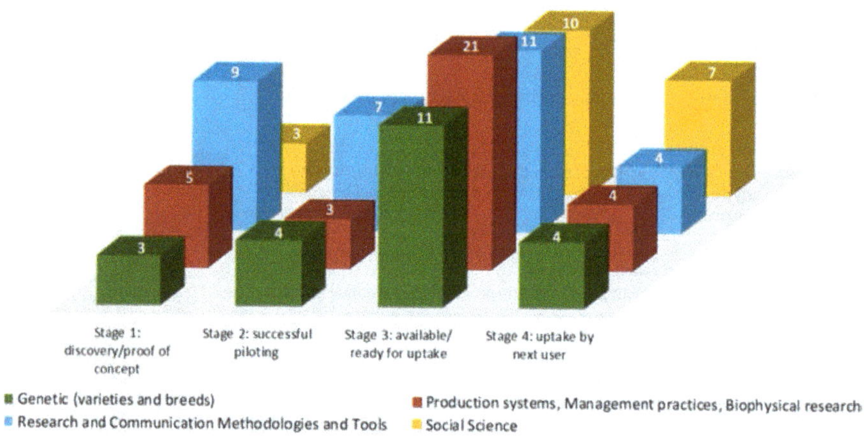

Fig. 2.14 Innovations in the RTB portfolio and their stage of readiness (2019)

aggregated innovations by type and level of readiness across the program and for each flagship (Figs. 2.14 and 2.15).

In summary, aggregated innovations made up the 25 clusters in the RTB portfolio in Phase 2. RTB's comprehensive framework for the portfolio also provided the basis for determining and linking groups of lead and complementary innovations. Aggregated innovations were embedded in a FP that included a theory of change with quantified indicators. Although RTB did not have sufficient resources to implement the M&E system as initially planned with all the cluster teams, where they were developed, co-constructed impact pathways helped to identify policy changes,

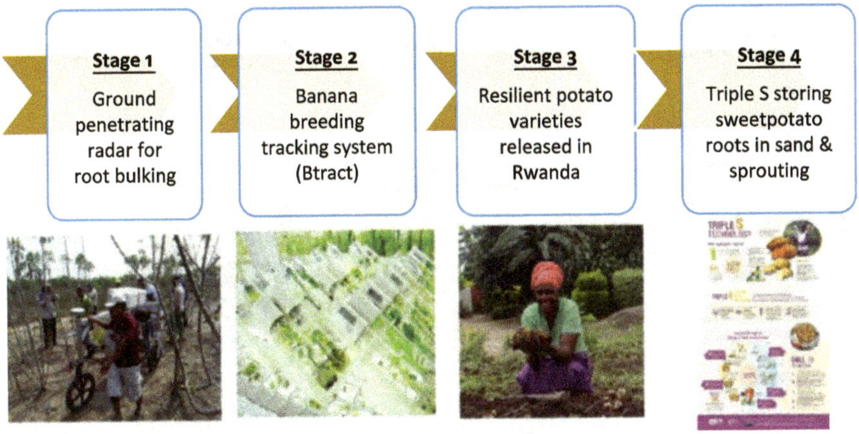

Fig. 2.15 FP 2: innovations reported (2017–2019)

outcomes, and milestones. The MEL software was effectively used in RTB monitoring and evaluation and adapted to better assess innovation readiness. This contributed to CGIAR-wide CRP reporting and performance assessment by the System Office.

2.5.5 *Programmatic Embedding of Strategic and Integrated Gender Research*

A key institutional innovation in RTB was to differentiate and embed two different, complementary types of research: (a) integrated gender research on specific technologies where gender scientists worked with biological and other social scientists to address specific gender constraints and opportunities for that technology and (b) strategic research addressing the knowledge gaps on how gender roles and norms affect the uptake of RTB technologies generally (Fig. 2.16).

During its first phase, RTB adopted a strategy to integrate gender in technical areas across the entire CRP and to conduct strategic gender research. The gender team comprised one gender specialist in each center, a gender coordinator based in the PMU, and part-time support from a senior gender specialist at CIP. The RTB gender team linked up with other gender scientists, in particular as key members of the CGIAR global study on Enabling Gender Equality in Agriculture and Natural Resource Management (GENNOVATE). RTB gender scientists wrote 15 case studies in target countries including Uganda (4), Malawi (2), Burundi (1), Nigeria (2), Colombia (4), Bangladesh (2), and Vietnam (2). GENNOVATE was a strategic research endeavor that explicitly sought to understand how household and community power relations and self-perceptions of personal power shape innovation decisions. A key learning from this strategic research was that when interventions do not

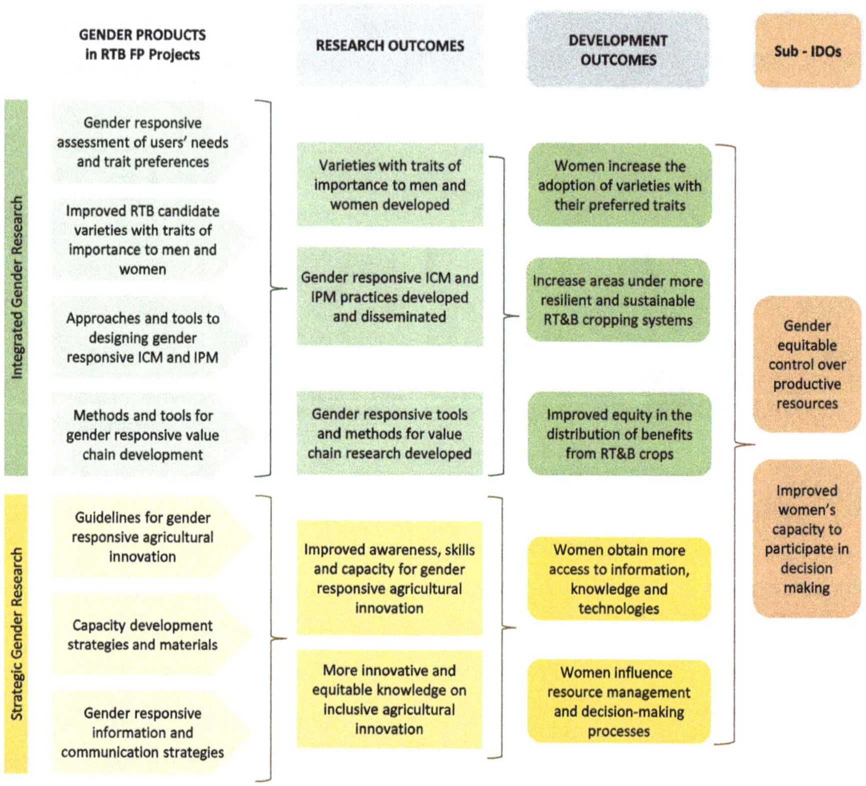

Fig. 2.16 Key elements from RTB gender impact pathway

consider underlying social structures and gender norms related to household decision-making and control of income, agriculture research may not benefit women.

Besides the strategic research in GENNOVATE, there was integrated gender research. For example, in Phase 1, a gender specialist in a seed potato project helped the team to realize that women did not have access to quality potato seed because they lacked access to credit and training and did not control household income to purchase seed (Mudege et al. 2015). Research on long shelf-life banana and potato ambient storage technology in Uganda reached similar conclusions.

During Phase 2, increased emphasis was given to integrated gender research supported by specific earmarked funding grants for gender research in some technological clusters. For example, in the breeding community of practice cluster, earmarked funding supported gender specialists and breeders to develop tools for gender-responsive breeding. Similarly, earmarked funding for gender research was provided to clusters working on seed systems and pest and disease management. By 2021, gender knowledge resources were organized in the Gender Responsive AR4D

Portal and shared with other CRPs its work on Gender Plus (G+) tools for gender-responsive breeding.

In summary, RTB adopted an approach to strengthen its gender research capacity, fostering feedback loops between strategic and integrated gender research. This required leadership of the RTB PMU, with support from the ISC, and earmarked funding directed to technological clusters and to the gender and youth cluster in FP5 that supported strategic research. This programmatic embedding generated concrete outputs to help address the challenges of gender and social inclusion for innovation design and scaling.

2.5.6 *Internal Funding Mechanisms and Incentives*

In Phase 2, total funding across all sources averaged around $80 million a year, making RTB the largest of all CRPs and the largest single program of any sort in the CGIAR. About 80% of RTB funding was managed directly by the centers (W3 and bilateral funding, or non-pooled funding[11]). Only about 20% was directly managed by RTB through contracts with CIP as the lead center (W1 and W2, or pooled funding[12]). This pooled funding was managed by the PMU, with the management committee and under the guidance of the ISC and, ultimately, the CIP Board of Trustees. Nevertheless, the proposals for both phases, the theories of change, and the milestones and deliverables were planned and reported for both pooled and non-pooled funding. Pooled funding was the smaller part of the overall budget, but it was carefully planned to add value and complement non-pooled funding to ensure the greatest impact.

Internal funding mechanisms for pooled funding evolved over time as PMU acquired more experience but broadly fell into three categories:

1. Earmarked funding (Table 2.5) represented about 31% of the budget allocated each year by RTB and was targeted to areas with the greater scope for synergistic value addition. This was related to multicenter investments that ordinarily were not covered through bilateral funding grants. Much of this was linked to multicenter work in the discovery and cross-cutting clusters. Additionally, some of the earmarked funding was linked to multicenter work in cassava and banana clusters where the mandate was shared between several centers. Each earmarked funding request was organized as a project with specific deliverables proposed in funding submission and then tracked over a 3-year period.

[9] Bilateral funding was where the donor did not cover the 2% charge made by the center to the system.

[10] Pooled funding was of two types: W1 funding, which was received from the system from donors contributing to all system entities with no restrictions, and W2 funding, which was awarded by donors specifically to RTB, sometimes with conditions on which flagship it should be used in.

Table 2.5 Earmarked funding awards in Phase 2 – 2017–2021

FP	Name of earmarked funding	Center/partners	Duration	Funding (USD, 000)
FP1	Breeding platform	Bioversity – CIAT – CIP – IITA	2017–2021	2184
FP1	Next-generation breeding	Bioversity – CIP – IITA – CIRAD	2017–2018	436
FP1	Game-changing traits	Bioversity – CIAT – CIP – IITA	2017–2021	1410
FP1	Genetic diversity	Bioversity – CIAT – CIP – IITA – CIRAD	2017–2021	1376
FP2	Quality seeds and access to improved varieties	Bioversity – CIAT – CIP – IITA – WUR	2017–2021	4343
FP2	GBI gender tool	IITA	2020	25
FP3	Pest/disease management	Bioversity – CIAT – CIP – IITA	2017–2021	3587
FP3	Crop production systems	Bioversity – CIAT – CIP – IITA	2017–2018	268
FP3	Banana fungal and bacterial diseases/ FOC/ BXW	Bioversity – CIAT – IITA – CIRAD	2017–2021	1355
FP3	Banana viral diseases/BBTV	Bioversity – IITA – CIRAD	2017–2021	1531
FP3	Cassava biological constraints, Asia/ Americas	CIAT – IITA	2019–2020	170
FP4	Postharvest innovation and nutrition improvement	Bioversity – CIAT – CIP – IITA	2017–2021	1128
FP4	Cassava processing	CIAT – IITA – CIRAD – NRI	2017–2021	1942
FP5	Foresight, impact, monitoring, and co-learning	Bioversity – CIAT – CIP – IITA – CIRAD – WUR	2017–2021	3663
FP5	Sustainable intensification and diversification for improved resilience, nutrition, and income	Bioversity – CIP – IITA – WUR	2017–2018	252
FP5	Gender equitable development and youth employment	Bioversity – CIP – IITA	2017–2021	1099
FP5	Institutional innovation, scaling	Bioversity – CIAT – CIP – IITA – CIRAD – WUR	2017–21	3075

2. Non-earmarked funding, about 57% of the annual budget, was available for centers to invest on specific deliverables in alignment with the proposal for the phase and theories of change and to complement Window 3 investments but was not projectized and therefore more flexibly managed.
3. Scaling funds, about 12% of the yearly budget, were allocated through competitive calls. These funds supported the scaling of innovations that had higher levels of readiness.

Earmarked funding was awarded and renewed through an internal competitive process with submissions of project proposals evaluated by the PMU based upon criteria developed collaboratively with the Management Committee and ISC. After feedback, many submissions received some funding, and the strongest ones received the most. This created a space for constrained competition, ensuring that most areas of synergistic investment received some funding.

During Phase 2, an institutional innovation called the RTB Scaling Fund was introduced to foster the scaling of innovations, generate an evidence base of their scalability, and improve scaling strategies (Sartas et al. 2020). As discussed above, scaling was a critical limitation identified in Phase 1. In 2017, RTB held a workshop with some stakeholders to review best bet scalable innovations. Later, RTB clusters with innovation packages in advanced stages of readiness were invited to apply for the RTB Scaling Fund. These awards were made for 2 years in order to move innovation packages to a higher level of scaling readiness. This funding was competitive with two external reviewers and one PMU member evaluating submissions against clearly defined criteria (Table 2.6).

The innovation packages receiving scaling fund support and the allocations are shown in Table 2.7.

In summary, RTB's use of W1, W2, and W3 funding mechanisms enabled maximum and strategic collaboration. The RTB Scaling Fund was a significant innovation to address scaling more comprehensively.

Table 2.6 Scaling Fund selection criteria (for concept notes and full proposals)

Criteria	Score Max (100)
1. Relevance Is the proposal relevant to the purpose of the fund? Is the contribution of the proposal in moving the innovation along the scaling readiness levels convincing? Are the objective and outcomes significant and realistic (number of actual beneficiaries already reached and number of expected beneficiaries in the coming 2 years)?	25
2. Partnerships Does the concept note provide solid elements to build the partnership and scaling strategies? Is the description of stakeholders' roles and synergies clear and convincing?	20
3. Scaling strategy Does the scaling strategy identify key opportunities and bottlenecks to achieve the project outcomes? Does the strategy clearly present how scaling of the proposed innovations will contribute to achieving project outcomes?	20
4. Strength of the proposed multidisciplinary team (it may include partners and staff funded through different sources)	15
5. Level of co-investment by key government, public or private scaling partners	20

Table 2.7 Scaling fund awards in Phase 2

Flagship	Innovation package	Center/ partners	Duration	Funding (USD, 000)
FP3/ FP5	Broadening the scaling of BXW management in East and Central Africa (Chap. 10)	Bioversity – IITA – WUR	2018–2019	700
FP2/ FP5	Scaling Sweetpotato Triple S PLUS – gender-responsive options for quality planting material, higher yields and extended shelf life for storage roots (Chap. 12)	CIP – WUR	2018–2019	701
FP4/ FP5	Scaling the transformation of wet cassava peels into high quality animal feed ingredients (Chap. 6)	IITA – CIAT – WUR	2018–2019	404
FP4/ FP5	Scaling approach for flash drying of cassava starch and flour at small scale (Chap. 4)	CIAT – IITA – CIRAD – WUR	2019–2020	903
FP2/ FP5	Market-driven scaling up and adoption of potato in Africa through a technology package combining climate resilient, novel potato varieties with a seed system innovation (Chap. 13)	CIP – WUR	2019–2020	1013
FP4/ FP5	Orange Fleshed Sweetpotato (OFSP) Puree for Safe and Nutritious Food Products and Economic Opportunities for Women and Youths in Kenya, Uganda and Malawi (Chap. 5)	Bioversity – CIP – WUR	2019–2020	977
FP3/ FP5	Scaling AKILIMO, a digital fertilizer recommendation service	CIP – IITA – WUR	2020–2021	1021
FP1/ FP5	Scaling RTB crop variety validation and diffusion using farmer citizen science in Ghana and Rwanda	Bioversity – CIP – IITA – WUR	2020–2021	1000

2.5.7 Dynamic Interactive Communication Capability to Build a Shared Vision and Engage Stakeholders

Using a strategic communication approach, RTB pursued shared goals that would not have been achievable if the centers had worked alone. RTB developed its goal-oriented strategy through a SWOT analysis in 2013 (RTB 2013a). The PMU hired a full-time specialist to develop the communication strategy, using social media campaigns and joint blogs and coordinating with focal points in the RTB centers. Building on the SWOT analysis, RTB communications served multiple functions, including the following:

(a) To advance the *image of RT&B crops* as relevant for investment and research for nutrition, food security, and women farmers' income generation. RTB communicated the importance of RT&B crops to policymakers, donors, and researchers.

(b) To promote the program externally as *a globally recognized leader on knowledge and research about RT&B crops* with its own branding. RTB developed branding guidelines, publication and acknowledgment guidelines, and an "about us" statement to ensure the brand was understood internally and could be differentiated from other CRPs and centers externally. RTB had its own website and published several blogs monthly about scientific achievements and development impact. RTB developed an illustrated annual report for stakeholders that complemented more technical reporting (CGIAR Research Program on Roots, Tubers and Bananas, 2021). This showcased the breadth of the program, its integrated vision, and specific focus on the collective assets and collaborative innovations. Communication products included videos and podcasts on social media (Facebook, Twitter, LinkedIn, YouTube). Activities and achievements were summarized in blogs and in a quarterly newsletter. RTB supported several international and regional events each year, in particular meetings of the International Society for Tropical Root Crops (ISTRC) where national partners participated.

(c) To support *internal communications* on RTB vision and goals. The communication function targeted RTB scientists, emphasizing why the program added value to what the centers and their scientists did and the demonstrated value of collaboration. RTB held community meetings to report progress and plan cross-cutting collaboration about every 18 months; these were organized around the thematic subthemes and then FPs. The RTB communication specialist supported networking among researchers and encouraged FP and cluster teams to communicate in the ways they were more comfortable with.

(d) *To identify and nurture collective knowledge assets*. As discussed in Sect. 2.5.6, RTB earmarked funding and the cross-cutting clusters that created multicenter communities of practice and generated collective knowledge assets. With the end of RTB in 2021 and the creation of new initiatives within One CGIAR, there was an urgent need to ensure that RTB's legacy and collective assets would find a place in future initiatives. The collective knowledge assets at risk in the transition were designated "golden eggs" (Fig. 2.17). The landing page of the RTB golden eggs and writing descriptions of each golden egg and promoting them through social media was a communications goal in 2021. The CAS Secretariat (2020) evaluation considered it critical for "RTB to develop and expand these packages to inclusively cover the full program achievements." The CGIAR as a whole subsequently adopted this legacy initiative.

In summary, RTB effectively used communications to manage program complexity by drawing on goal-oriented internal and external communications and using communications strategically to draw attention to RT&B crops. As an institutional innovation, the RTB collective knowledge assets or golden eggs identify the legacy of the RTB CRP, and the close link between scaling innovations and communications informs new initiatives in One CGIAR.

Fig. 2.17 RTB golden eggs, key collective knowledge assets

2.5.8 Purposive National Partner Engagement

RTB works with national partners, especially within the national agricultural research systems (NARS) through principal program participants, and the CGIAR centers. Consequently, RTB's higher-level, coordinating role is not always visible to those partners. Nevertheless, RTB had a strategy for engaging national partners, for instance, during stakeholder consultations and in some cases while preparing theories of change. However, RTB also opted to use existing events where national partners came together, rather than creating its own dedicated partnership platforms. RTB supported three such partnership platforms, the most important of which was the ISTRC.

RTB supported both the general meeting of the ISTRC and of its Africa branch (ISTRC-AB). These meetings brought ownership of and participation by national scientists and extensionists, with an increasing participation of African members. The ISTRC meetings came the closest to a generalized platform for the RT&B crops, as all are included with the exception of bananas. RTB also co-organized its own meetings to follow up on these events. This strengthened the ISTRC meetings by enhancing attendance and knowledge sharing. On multiple occasions, RTB organized presentations of the program and of each flagship on the final day of ISTRC meetings, preceding the RTB meeting, where national scientists shared their experiences. RTB also engaged national partners at the Africa Potato Association and the Global Cassava Partnership for the 21st Century. For all these meetings, RTB made small travel grants to encourage young female and male scientists to present their work. This national partner engagement created a broader awareness of RTB's contributions and its role in partnerships and advocacy on RT&B crops.

RTB also had two experiences where the PMU directly managed projects in Uganda with the RTB-ENDURE project (Bentley et al. 2021) and in Nigeria with BASICS (Bentley et al. 2020a, b). These two experiences created deeper collaborative networks with national partners in these two countries and helped to capitalize on experiences from the wider RTB community to benefit national partners.

2.6 Program Outcomes

RTB's institutional innovations facilitated various outcomes across centers (Table 2.8). These outcomes are concentrated in the areas of breeding, seed systems, pests and diseases, and postharvest, all of which were built on significant bilateral investment that complemented the earmarked funding for cross-center collaboration. Agronomy was a high priority for RTB, but in the absence of significant bilateral investment, its cross-cutting outcomes were limited. This changed in 2019 with the cassava agronomy decision support tool AKILIMO (https://www.akilimo.org/) led by IITA. This digital application and database could incorporate other RTB crops and therefore scale to other contexts and uses. Table 2.8 does not include other notable center-specific outcomes such as breeding and scaling of the orange-fleshed sweetpotato (Low and Thiele 2020).

As discussed in Sect. 2.5, the RTB 2.0 portfolio identified impact pathways that could be linked directly to the Intermediate Development Outcomes (IDOs) as set out in the CGIAR Strategic Research Framework (SRF). Figure 2.18 shows how RTB tracked contributions to system-level outcomes linked to the United Nations' Sustainable Development Goals, particularly SDG1, SDG2, and SDG13.

2.7 Lessons Learned from RTB

One of the difficulties of a complex program like RTB is to grasp how the partnership contributed to research and development outcomes and to document evidence of value at scale (Horton et al. 2009). RTB's complex institutional innovations (Sect. 2.5) facilitated the outcomes presented in Table 2.8 and Fig. 2.16. RTB made value at scale possible because there was:

- A common, shared vision and purpose and realistically defined goals
- Support for the partnership from participating organizations
- Equitable sharing of resources, responsibilities, and benefits
- Transparent governance and decision-making
- Creation of genuine respect and trust between the partners
- Achievement of higher-level outcomes beyond the partnership itself
- Committed leadership in the RTB Management Committee and PMU, notably the role of the Program Director with deep RTB knowledge and many years of experience working in the CGIAR and its partnership networks.

RTB was considered a strong model of good partnering within the CGIAR and by its stakeholders with transparent and equitable decisions about the use of funding mechanisms and the program direction (CGIAR-IEA 2016; ISPC 2016; CAS Secretariat 2020). Collaboration and partnering were articulated with an effective structuring of its portfolio around cross-cutting clusters or innovation packages that facilitated reciprocal learning exchange across crops and among partners. This

Table 2.8 Eight key RTB research outcomes based upon multicenter collaboration

Cluster/partners	Research outcome	How/when
DI1.1 Breeding community of practice, with Excellence in Breeding (EiB)	Researchers use more focused design of breeding products to meet farmer and consumer demands and improved management of breeding product pipelines	2013: RTB brought breeders together 2016: Creation of a Breeding Community of Practice with earmarked funding, comparing breeding strategies and developing shared methods, including the Tricot citizen science approach for varietal testing 2019: Product profiles registered (47) for the main targets of RTB breeding registered with CGIAR EiB Platform 2020: Tools to incorporate gender into product profile development are tested and adapted in collaboration with EiB 2020: Hackathon, breeders, social scientists, gender specialists, and food scientists peer-review and improve market segment definitions and variety product profiles for four breeding programs
DI1.2 Next-generation breeding, with Royal Holloway, University of London (RHUL)	Scientists at IITA, CIAT, and CIP incorporate design, metabolite extraction, and interpretation and use metabolomic data for all crops	2013: Theme 2 leader (CIAT) puts together an earmarked funding proposal across all centers and crops to begin this work 2018–2020: Ten peer-reviewed publications published; see overview in Friedmann et al. (2019) 2020: Compound database and concentration range for metabolites detected in the major RT&B crops available for breeding programs
CC2.1 Seed systems (with WUR and U Florida)[a]	Improved seed systems. Seed system toolbox validated. Thirteen tools developed and web accessible, for improving the design and execution of seed-system interventions and the management of seed degeneration	2014: Researchers from CGIAR centers, Wageningen University & Research (WUR), and Kansas State University completed and analyzed 12 case studies of RT&B seed systems in Africa and Latin America using common framework (RTB 2016b). Improved models for seed degeneration management developed based on field trials and integrated seed health strategy. 2016: Community of Practice Cluster CC2.1 funded with earmarked funds 2017: Cluster CC2.1 collaborates to incorporate tools for understanding RTB seed systems into a single toolbox using a standard format. 2020: Toolbox validated in 14 projects in Asia, Africa, and South America across all major RTB crops 2021: Online version of the Toolbox available for use by government agencies, NARS, NGOs, and donors

(continued)

Table 2.8 (continued)

Cluster/partners	Research outcome	How/when
CC3.1 Pests and diseases (with various NARS and ARIs)[b]	Researchers, agricultural ministry officials, national plant protection organizations, and extension agents use tools and strategies to manage major pests and diseases	2012: Joint RTB and CCAFS workshop on management of critical pests and diseases through enhanced risk assessment and surveillance and understanding of climate impacts through enhanced modeling 2015: In a cross-crop consultative intervention of all RTB centers, key regional target pests and diseases were identified for pest risk analysis (PRA) 2017: Cassava Disease Surveillance (CDS) virtual network supports accurate diagnosis and offers solutions for prevention and management. CGIAR and national partners strengthen capacities to perform PRAs predicting risk of insect-transmitted viruses and generation of georeferenced risk maps 2018: ICT tools used to identify major diseases in the field for surveillance and plant disease management 2020: Digital alliance for pests and diseases as a golden egg RTB ensures gender is addressed across topics in the webinars for the International Year of Plant Health
CA3.5 Cassava biological constraints, Asia/ Latin America, with 21 NARS, national plant protection organizations (NPPOs, U of Queensland)	Scientists use response plan developed to contain emerging cassava mosaic disease (CMD) in SE Asia	2015: CMD detected in Cambodia. Network of experts to monitor and manage cassava mealybug and cassava witches' broom disease in Asia, progress in helping farmers combat pests, with local partners 2018: Regional workshop supported by the Global Cassava Partnership for the Twenty-first Century and CIAT led to the joint development of a response plan, with IITA participation under the RTB umbrella 2018: Adoption of biological control for cassava mealybug using a host-specific parasitoid provided by IITA involving collaboration from CIAT, contributed to restoring the cassava yield that had dropped by 27% after the arrival of the cassava mealybug (first reported in 2008) 2019: Improved capacities of NPPOs and cassava farmers to contain CMD and implement management strategies in Cambodia, Laos, Thailand, and Vietnam following a multi-pronged approach (Siriwan et al. 2020)

(continued)

Table 2.8 (continued)

Cluster/partners	Research outcome	How/when
BA3.4 Banana viral diseases (with 14 NARS, FAO, U Queensland)	Multinational, multi-stakeholder Alliance coordinated action to halt the expansion of banana bunchy top disease (BBTD) and recover banana production in disease-affected areas, especially in sub-Saharan Africa	2014: RTB learning alliance to contain BBTD and help farmers recover from it in eight African countries; established field sites and capacity building for researchers from each participating country 2015: BBTD Alliance launched by RTB. Fifteen pilot sites in eight African countries set up to implement strategies for recovering banana production (https://www.bbtvalliance.org/) 2019: Training workshops in Togo, Nigeria, and Cameroon 2021: Gender roles analysis (Nkengla-Asi et al. 2021) 2021: Alliance developing capacity for disease recognition and control options, training farmers and entrepreneurs in field inspection, rogueing infected plants and production of clean planting material 2021: Training African scientists to continue surveillance on the extent of disease incidence to update the spatial distribution map of BBTV spread in SSA 2021: Location-specific clean banana production and distribution systems are being established to replenish banana plants that have been eradicated
CC4.1 Postharvest innovation, led by CIRAD with eight NARS and Cornell U	Framework and tools to breed for quality traits in 11 food products from RTB crops in five target countries	2015 RTB supported work on end-user preferences through initiatives with national partners and the Natural Resources Institute (NRI, UK) 2018: RTBfoods project funded by Bill & Melinda Gates Foundation (BMGF) (https://rtbfoods.cirad.fr/) 2020–2021: Special journal issue (Dufour et al. 2021) published with 17 articles on end-user preferences of RT&B crops and method to measure quality traits

(continued)

Table 2.8 (continued)

Cluster/partners	Research outcome	How/when
CC5.4 Scaling, with WUR[c]	Scaling readiness approach implemented across the RTB portfolio, in collaboration with the scaling experts of CC5.4	2017: Scaling Readiness Approach and RTB Scaling Fund developed and piloted 2018–2021: Eight projects supported by RTB Scaling Fund 2019: Scaling Readiness web portal (www.scalingreadiness.org) and principles (Sartas et al. 2020) published 2020–2021: Other CRPs, CGIAR centers, and the CGIAR System Office expressed interest in RTB's scaling work and using the Scaling Readiness approach 2021: RTB gender research, tools, and methods are compiled and shared for broader use in collaboration with the Gender Platform

[a]See Chap. 11 of this book
[b]See Chap. 9 of this book
[c]See Chap. 3 of this book

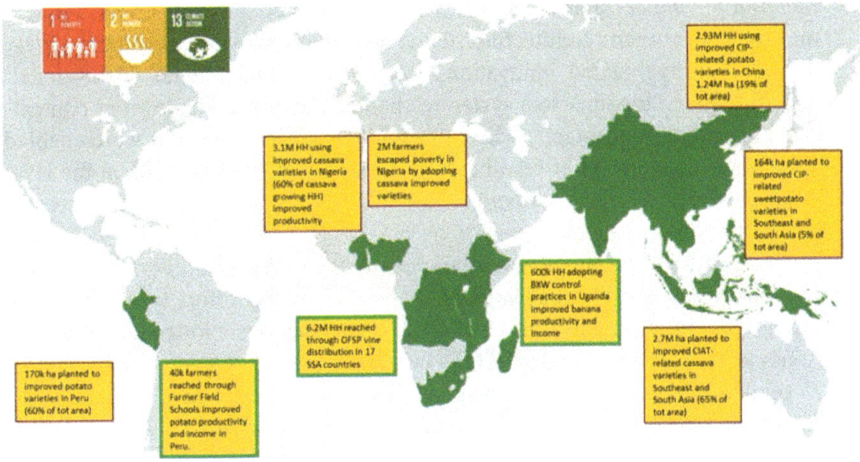

Fig. 2.18 RTB's contributions to system level outcomes, 2017–2020.

value addition by RTB aligned with institutional innovation around scaling that was a strong feature of RTB 2.0.

One challenge with RTB closing at the end of 2021 is that the organizing principles of the new initiatives in the One CGIAR (2022–2030) are not organized by crop type (CGIAR System Organization 2021). There should be significant work continuing on the RT&B crops, but the RTB program and its clusters will be redistributed and work will be organized in different ways. The RTB golden eggs provide some continuity of innovation packages, but not necessarily the synergies among the teams. At the time of writing, there is interest in keeping the RT&B crops together as a work package in the One CGIAR seed initiative, but it is less clear what happens to other RTB work in One CGIAR.

In summary, various lessons emerge from the RTB experience that are relevant for future multi-organizational programs for agricultural research for development and for One CGIAR.

(a) *Build on the unique value proposition of the collaborative program.* For RTB, this relates to the commonalities of clonally propagated crops, with program design to add value to them through the cross-cutting clusters and their communities of practice on topics of breeding, seed systems, pests and diseases, and postharvest, all of which had dedicated funding. A recommendation for One CGIAR would be to keep clonally propagated crops together as a group within key research areas to further enhance synergies and achievements.
(b) *Use strategic communications to build broad-based stakeholder support around the value proposition.* RTB did this with stakeholder consultations and communications demonstrating progress.
(c) *Ensure that programmatic design can bring together the different innovations required for impact.* For RTB, this happened at three different levels. First, at the level of program architecture, the crop-specific clusters were purposively designed as aggregated innovations (technical, organizational, institutional) and contributed together to a systemic change. Second, at the level of context-specific scaling fund grants, where specific innovation packages were enabled to move to a high level of readiness and greater impact. And third, at the level of purposively engaged partners who could scale results further and created further systemic change.
(d) *Embed gender in technical research and engage with and support researchers to mainstream gender in biological sciences.* Without effort to resource and monitor the progress of gender in A4RD, the impact of socio-technical innovations can be limited or eroded. RTB's targeted support helped to achieve impact in gender in breeding and the uptake of G+ tools for breeding.
(e) *Establish structural incentives for integration among initiatives and across crops* if cross-CGIAR contributions to the IDOs and SDGs are to be fully captured. This was an area where RTB and other CRPs underachieved compared to the promise at program design. One of the primary reasons for this was insecurity and late award of funding, which made it more challenging to secure strategic partnerships.
(f) *Link theories of change with flexible and utilization-focused M&E systems* to strengthen adaptive management and reflexivity at different management levels. This can enhance planning and reporting, including milestones and indicators used to plan and report at the level of the CGIAR system, which were excessively rigid for CRPs.

(g) *Recognize and incorporate key partnerships* that add value to the program participants. These need to respect good partnering principles including identification of the value added from the partnership, transparency and access to funding, shared responsibilities, and clarity of the role of independent advisory groups to adjudicate any conflicts in the partnership. When the program is closed, attention needs to be paid to ensure that relationships (corporate, technical, and personal) built up by RTB and other CRPs are retained. For this reason, this book intentionally documents RTB experiences.
(h) *Develop mechanisms to promote continuity of key innovations and teams of collaborators associated with them (inside and outside of CGIAR) as the program is closed.* RTB, for example, used the concept of golden eggs with resourced and linked nurturing plans.

2.8 Conclusion

RTB demonstrates that institutional innovations involving collaborative program design and management can enable comprehensive research and development outcomes aligned to program goals. Over its 10 years, RTB evolved as a unique, global collaborative program with connections among multiple centers and stakeholders bridging upstream research, translational research, and innovation processes to deliver value at scale. RTB developed pragmatic institutional innovations to manage the program effectively, ensuring research outcomes and contributing to the global Sustainable Development Goals. Being embedded within the CGIAR has required RTB to adapt its capacity to respond to system change. One CGIAR could draw on the RTB approach to collaboration as complex institutional innovations organized by a portfolio of aggregated innovations that led to concrete achievements. These innovations include the strategic use of funding mechanisms, co-constructed impact pathways, stakeholder consultations and priority setting, goal-oriented communications, committed and effective program leadership, and governance and partnerships for strategic interventions to realize cross-cutting priorities, such as strategic and integrated gender research. The following chapters in this book describe RTB's legacy in greater detail.

Acknowledgments The research reported here was undertaken as part of, and funded by, the CGIAR Research Program on Roots, Tubers and Bananas (RTB) and supported by CGIAR Trust Fund contributors (https://www.cgiar.org/funders/) as well as numerous other donors.

The authors thank the following individuals who provided input and feedback on earlier drafts of this chapter: Rupert Best, Rodney Cooke, Jill Lenné, Gordon Prain, Oscar Ortiz, Selcuk Ozgediz, and Yvonne Pinto. We greatly appreciate Zandra Vasquez's assistance with references and formatting.

References

Almekinders CJM, Walsh S, Jacobsen KS, Andrade-Piedra JL, McEwan MA, de Haan S, Kumar L, Staver C (2019) Why interventions in the seed systems of roots, tubers and bananas crops do not reach their full potential. Food Security 11:23–42. https://doi.org/10.1007/s12571-018-0874-4

Arena M, Cross R, Sims J, Uhl-Bien M (2017) How to capitalize innovation in your organization. MIT Sloan Manag Rev. https://sloanreview.mit.edu/article/how-to-catalyze-innovation-in-your-organization

Barrett CB, Benton Y, Fanzo J, Herrero M, Nelson RJ, Bageant E, Buckler E, Cooper K, Culotta I, Fan S, Gandhi R, James S, Kahn M, Lawson-Lartego L, Liu J, Marshall Q, Mason-D'Croz D, Mathys A, Mathys C, Mazariegos-Anastassiou V, Miller A, Misra K, Mude AG, Shen J, Majele Sibanda L, Song C, Steiner R, Thornton P, Wood S (2020) Socio-technical innovation bundles for agri-food systems transformation, report of the international expert panel on innovations to build sustainable, equitable, inclusive food value chains. Cornell Atkinson Center for Sustainability and Springer Nature, Ithaca and London

Baum WC, Lejuene M (1986) Partners against hunger: the consultative group on international agricultural research. The World Bank for the CGIAR, Washington, DC

Bentley J, Nitturkar H, Friedmann M, Thiele G (2020a) BASICS phase I final report. 23 Dec 2020. 78

Bentley J, Nitturkar H, Friedmann M, Thiele G (2020b) Is there a space for medium-sized cassava seed growers in Nigeria? J Crop Improv. https://doi.org/10.1080/15427528.2020.1778149p

Bentley JW, Naziri D, Prain G, Kikulwe E, Mayanja S, Devaux A, Thiele G (2021) Managing complexity and uncertainty in agricultural innovation through adaptive project design and implementation. Dev Pract 31(2):198–213. https://doi.org/10.1080/09614524.2020.1832047

Berthet ET, Hickey GM, Klerkx L (2018) Opening design and innovation processes in agriculture: insights from design and management sciences and future directions. Agric Syst 165:111–115

Byerlee D, Lynam J (2020) The development of the international center model for agricultural research: a prehistory of the CGIAR. World Dev 135:105080. https://doi.org/10.1016/j.worlddev.2020.105080

CAS Secretariat (CGIAR Advisory Services Shared Secretariat) (2020) CGIAR research program 2020 reviews: roots, tubes and bananas (RTB). CAS Secretariat Evaluation Function, Rome. https://cas.cgiar.org/evaluation/publications/crp-2020-review-roots-tubers-and-bananas-rtb

CGIAR (2015a) Research program portfolio report 2015 https://hdl.handle.net/10947/4480

CGIAR (2015b) CGIAR strategy and results framework 2016–2030: overview. https://hdl.handle.net/10947/4069

CGIAR (2016) Research program portfolio report 2016 https://hdl.handle.net/10568/89018

CGIAR Research Program on Roots, Tubers and Bananas (2018) RTB Annual report 2017. Phase II – year 1. Lima (Peru). https://hdl.handle.net/10568/97476

CGIAR Research Program on Roots, Tubers and Bananas (2019) RTB Annual report 2018. Phase II – year 2. Lima (Peru). https://hdl.handle.net/10568/103623

CGIAR Research Program on Roots, Tubers and Bananas (2020) RTB Annual report 2019. Phase II – year 3. Lima (Peru). https://hdl.handle.net/10568/108931

CGIAR Research Program on Roots, Tubers and Bananas (2021) RTB Annual report 2020: innovation and impact. Phase II – year 4. Lima (Peru). https://cgspace.cgiar.org/handle/10568/114576

CGIAR System Organization (2021) CGIAR 2030 research and innovation strategy: transforming food, land, and water systems in a climate crisis. CGIAR System Organization, Montpellier. https://hdl.handle.net/10568/110918

CGIAR-IEA (2015a) Evaluation of CGIAR Research Program on Roots, Tubers and Bananas (RTB), vol I. Independent Evaluation Arrangement (IEA) of the CGIAR, Rome. http://iea.cgiar.org/

CGIAR-IEA (2015b) Evaluation of CGIAR Research Program on Roots, Tubers and Bananas (RTB), vol II. Independent Evaluation Arrangement (IEA) of the CGIAR, Rome. http://iea.cgiar.org/

CGIAR-IEA (2016) CGIAR Research Program on Roots, Tubers, and Bananas. Evaluation brief. https://cas.cgiar.org/sites/default/files/pdf/Evaluation-Brief-RTB.pdf

Coffman WR, Acevedo M, Evanega SD, Porciello J, Tufan HA, McCandless L (2020) Viewpoint: five recommendations for an inclusive and collaborative One CGIAR. Food Policy 91:101831. https://doi.org/10.1016/j.foodpol.2020.101831

Consortium Board (2011) A strategy and results framework for the CGIAR. https://storage.googleapis.com/cgiarorg/2011/08/CGIAR-SRF-Feb_20_2011.pdf

CRP-Humidtropics (2016) Annual performance report. https://cgspace.cgiar.org/bitstream/handle/10568/89312/HUMIDTROPICS-Annual-Report-2016.pdf?sequence=1&isAllowed=y

CRP-RTB (2011) CRP-RTB 3.4. Roots, tubers, and bananas for food security and income. Revised proposal. 170 p. https://hdl.handle.net/10568/83375

Dalrymple D (2008) International agricultural research as a global public good: concepts, the CGIAR experience and policy issues. J Int Dev 20:347–379. https://doi.org/10.1002/jid.1420

Dufour D, Hershey C, Hamaker B, Lorenzen J (2021) Integrating end-user preferences into breeding programmes for roots, tubers and bananas. Int J Food Sci Technol 56(3):1071–1075. https://doi.org/10.1111/ijfs.14911

Ekboir J (2009) The CGIAR at a crossroads: assessing the role of international agricultural research in poverty alleviation from an innovation systems perspective. ILAC Working Paper 9, Institutional Learning and Change Initiative, Rome

Feldman S, Biggs S (2012) International shifts in agricultural debates and practice: an historical view of analyses of global agriculture. In: Campbell W, López Ortíz S (eds) Integrating agriculture, conservation and ecotourism: societal influences. Issues in agroecology – present status and future prospectus, vol 2. Springer, Dordrecht. https://doi.org/10.1007/978-94-007-4485-1_2

Friedmann M, Becerra LA, Fraser PD (2019) Metabolomics in CGIAR Research Program on Roots, Tubers and Bananas (RTB). Lima (Peru). CRP RTB. RTB working paper 2019-2. ISSN 2309-6586. 11 p. https://doi.org/10.4160/23096586RTBWP20192

Fuglie K, Thiele G (2009) Research priority assessment at the international potato center (CIP). In: Raitzer DA, Norton GW (eds) Prioritizing agricultural research for development. CABI, Cambridge

GCARD (2011) The GCARD road map: transforming agricultural research for development (AR4D) systems for global impact. http://www.fao.org/docs/eims/upload/290017/the_gcard_road_map_finalized.pdf

Glaeser B (1987) The green revolution revisited. Allen & Unwin, London

Greenland DJ (1997) International agricultural research and the CGIAR system- past, present, and future. J Int Dev 9(4):459–482

Hardin LS, Collins NR (1974) International agricultural research: organising themes and issues. Agric Adm 1(1):13–22. https://doi.org/10.1016/0309-586X(74)90024-7

Harwood R, Kassam AH (2003) Research towards integrated natural resources management. Examples of research problems, approaches and partnerships in action in the Rome: CGIAR. Consultative Group on International Agricultural Research (CGIAR); Food and Agriculture Organization of the United Nations (FAO), 168 p

Horton D, Prain G, Thiele G (2009) Perspectives on partnership: a literature review. Working paper 2009-3. CIP, Lima, p 111

Immonen S, Cooksy L (2019) Role and use of independent evaluation in development-oriented agricultural research: the case of CGIAR, an agricultural research network. Outlook Agric 48(2):94–104. https://doi.org/10.1177/0030727019850835

Interim Science Council (TAC) (2003) Eighty-fourth meeting, June 9–13, 2003: end of meeting report. https://hdl.handle.net/10947/654

ISPC (2012) Strategic overview of CGIAR research programs, Part I. Theories of change and impact pathways. Independent Science and Partnership Council, CGIAR, Rome

ISPC (2016) ISPC assessment of the roots, tubers and bananas (RTB) CRP II revised proposal (2017–2022). Independent Science and Partnership Council, CGIAR, Rome

Krishna Bahadur KC, Dias GM, Veeramani A, Swanton CJ, Fraser D, Steinke D et al (2018) When too much isn't enough: does current food production meet global nutritional needs? PLoS One:16. https://doi.org/10.1371/journal.pone.0205683

Low J, Thiele G (2020) This understanding innovation: the development and scaling of orange-fleshed sweetpotato in major African food systems. Agric Syst 179(102770). https://doi.org/10.1016/j.agsy.2019.102770

McCalla AF (2014) CGIAR reform – why so difficult? Review, reform, renewal, restructuring, reform again and then "The New CGIAR" – so much talk and so little basic structural change-why? Working paper no. 14-001. UC Davis. p 55

McCalla AF (2017) The relevance of the CGIAR in a modernizing world. In: Pingali P, Feder G (eds) Agriculture and rural development in a globalizing world: challenges and opportunities, 1st edn. Routledge. https://doi.org/10.4324/9781315314051

Mudege NN, Kapalasa E, Chevo T, Nyekanyeka T, Demo P (2015) Gender norms and the marketing of seeds and ware potatoes in Malawi. J Gender Agric Food Secur 1(2):18–41

Nkengla-Asi L, Eforuoku F, Olaosebikan O, Ladigbolu TA, Amah D, Hanna R, Kumar PL (2021) Gender roles in sourcing and sharing of banana planting material in communities with and without banana bunchy top disease in Nigeria. Sustainability 13(6):3310. https://doi.org/10.3390/su13063310

Operations Evaluation Department (2004) The CGIAR at 31: an independent meta-evaluation of the Consultative Group on International Agricultural Research. Revised Edition. The World Bank, Washington, DC. http://hdl.handle.net/10986/15041

Ozgediz S (2012) The CGIAR at 40. Institutional evolution of the world's premier agricultural research network. CGIAR Fund Office, Washington, DC. http://hdl.handle.net/10986/23845

Pant LP, Hambly HO (2009) The promise of positive deviants: bridging divides between scientific research and local practices in smallholder agriculture. Knowl Manag Dev J. https://doi.org/10.1080/18716340903201504

Pemsl DE, Staver C, Hareau G, Alene AD, Abdoulaye T, Kleinwechter U, Labarta R, Thiele G (2022) Prioritizing international agricultural research investments: lessons from a global multi-crop assessment. Res Policy 51(4):104473. https://doi.org/10.1016/j.respol.2022.104473

Pigford AE, Hickey G, Klerkx L (2018) Beyond agricultural innovation systems? Exploring an agricultural innovation ecosystems approach for niche design and development in sustainability transitions. Agric Syst 164:116–121. https://doi.org/10.1016/j.agsy.2018.04.007

Price EJ, Drapal M, Perez-Fons L, Amah D, Heider B, Rouard M, Swennen R, Becerra Lopez-Lavalle LA, Fraser PD (2020) Metabolite database for root, tuber and banana crops to facilitate modern breeding in understudied crops. Plant J 103(5):1959–1959. https://doi.org/10.1111/tpj.14649

Renkow M, Byerlee D (2010) The impacts of CGIAR research: a review of recent evidence. Food Policy 35:391–402. https://doi.org/10.1016/j.foodpol.2010.04.006

Roberts N, Bradley R (1991) Stakeholder collaboration and innovation: a study of public policy initiation at the state level. J Appl Behav Sci 27:209–227. https://doi.org/10.1177/0021886391272004

RTB (2013a) RTB media and communication strategy for 2013–2014. CIP, Lima

RTB (2013b) Planning for greater impact: RTB current thinking 2016–2024. CIP, Lima

RTB (2016a) CGIAR research program on root, tubers and bananas. 2016. Full proposal 2017–2022. CIP, Lima. http://hdl.handle.net/10945/53701

RTB (2016b) User's guide to multistakeholder framework for intervening in RTB systems. CGIAR Research Program on Roots, Tubers and Bananas RTB Working Paper 2016–1

Sartas M, Schut M, Proietti C, Thiele G, Leeuwis C (2020) Scaling readiness: science and practice of an approach to enhance the impact of research for development. Agric Syst 183(102874)

Scott G (2020) A review of root, tuber and banana crops in developing countries: past, present, and future. Int J Food Sci Technol. https://doi.org/10.1111/ijfs.14778

Siriwan W, Jimenez J, Hemniam N, Saokham K, Lopez-Alvarez D, Leiva AM, Martinez A, Mwanzia L, Becerra Lopez-Lavalle LA, Cuellar WJ (2020) Surveillance and diagnostics of the emergent Sri Lankan cassava mosaic virus (Fam. Geminiviridae) in Southeast Asia. Virus Res 285:197959., ISSN 0168-1702. https://doi.org/10.1016/j.virusres.2020.197959

Woodhill J (2010) Capacities for institutional innovation: a complexity perspective. IDS Bull 41:47–59. https://doi.org/10.1111/j.1759-5436.2010.00136.x

Woolley J, Johnson VB, Ospina B, Lemaga B, Jordan T, Harrison G, Thiele G (2011) Incorporating stakeholder perspectives in international agricultural research: the case of the CGIAR Research Program for Roots, Tubers, and Bananas for Food Security and Income. International Potato Center (CIP), Lima, Peru. Social sciences working paper 2011-3. p 92

3

CGIAR Research on RTB: Scaling Readiness Approach

Marc Schut, Cees Leeuwis, Murat Sartas, Luis Alejandro Taborda Andrade, Jacob van Etten, Anna Muller, Thierry Tran, Arnaud Chapuis, and Graham Thiele

Abstract Scaling of innovations is a key requirement for addressing societal challenges in sectors such as agriculture, but research for development programs struggles to make innovations *go to scale*. There is a gap between new complexity-aware

M. Schut (✉) · M. Sartas
Wageningen University, Wageningen, the Netherlands

International Institute of Tropical Agriculture (IITA), Kigali, Rwanda
e-mail: marc.schut@wur.nl; m.schut@cgiar.org; murat.sartas@wur.nl

C. Leeuwis
Wageningen University, Wageningen, the Netherlands
e-mail: cees.leeuwis@wur.nl

L. A. Taborda Andrade
Centro Internacional de Agricultura Tropical (CIAT), Cali, Colombia

Universidad Nacional de Colombia, Bogotá, Colombia
e-mail: latabordaa@unal.edu.co

J. van Etten · A. Muller
Bioversity International, Rome, Italy
e-mail: j.vanetten@cgiar.org; a.muller@cgiar.org

T. Tran
Alliance Bioversity-CIAT, Cali, Colombia

CIRAD UMR Qualisud, Montpellier, France
e-mail: thierry.tran@cirad.fr

A. Chapuis
CIRAD, Montpellier, France
e-mail: arnaud.chapuis@cirad.fr

G. Thiele
CGIAR Research Program on Roots, Tubers and Bananas (RTB), led by the International Potato Center, Lima, Peru
e-mail: g.thiele@cgiar.org

scientific theories and perspectives on innovation and practical approaches that can improve strategic and operational decision-making in research for development interventions that aim to scale innovations. To bridge this gap, Scaling Readiness was developed. Scaling Readiness is an approach that encourages critical reflection on how ready innovations are for scaling in a particular context for achieving a particular goal and what appropriate actions could accelerate or enhance scaling to realize development outcomes. Scaling Readiness provides decision support for (1) characterizing the innovation and innovation system; (2) diagnosing the current readiness and use of innovations; (3) developing strategies to overcome bottlenecks for scaling; (4) facilitating multi-stakeholder negotiation and agreement; and (5) navigating the implementation process. This chapter explains how Scaling Readiness was used in the CGIAR Research Program on Roots, Tubers and Bananas (RTB) and describes how Scaling Readiness informed the design and management of the RTB Scaling Fund, an instrument for identifying and nurturing scaling-ready innovations. We introduce the key principles and concepts of Scaling Readiness and provide a case study of how Scaling Readiness was applied for scaling a cassava flash dryer innovation in different countries in Africa and Central America. The chapter concludes with a reflection and recommendations for the further improvement and use of Scaling Readiness.

3.1 Scaling of Innovation and Scaling Readiness

Innovation and the use of innovations at scale form an important element for achieving the Sustainable Development Goals (SDGs). The international community invests significant resources in the design and testing of innovations to overcome key challenges such as food insecurity, malnutrition, and environmental degradation. Many of those innovations start as bright ideas that are consequently developed in controlled conditions, to be tested with farmers and other end users. Although initial results and testing are often promising, relatively few lead to the desired positive impact at scale (Woltering et al. 2019). The agricultural research for development (AR4D) sector, in which the CGIAR is an important player, has been struggling with the question of how to best nurture impactful innovation and scaling pathways (Leeuwis et al. 2018).

One of the main reasons why innovations do not lead to impact at scale is that ideas about how scaling happens are not realistic. The notion of "find out what works and do more of the same" (Wigboldus et al. 2016) does not take into account the complex and diverse biophysical, socioeconomic, and political contexts that shape agriculture across the globe and limit the effectiveness of one-size-fits-all approaches (Hammond et al. 2020). Furthermore, research organizations often focus on technological innovations and pay less attention to the behavioral, organizational, and institutional changes that are needed to enable the effective use of technology (Schut et al. 2016). In addition, we observe that R4D interventions often see scaling as something that happens at the end of a short-term project or program as opposed to long-term systemic change processes and that scientists have limited capacities to shape impactful processes and partnerships needed for the scaling of innovations (Schut et al. 2020).

Revealing misconceptions and bottlenecks to the scaling of innovations, and supporting the development and implementation of effective strategies to overcome them, inspired the development of Scaling Readiness. In this chapter, the term "Scaling Readiness" (capitalized) is used as a brand name for the decision-support process that we have developed and as a key concept and metric that scores the maturity and scalability of an innovation (not capitalized).

The notion of "readiness" refers to whether an innovation has been tested and validated for the role it is intended to play in a specific context. The concept resonates with levels of technology readiness that have been proposed by the National Aeronautics and Space Administration (NASA) of the United States, the European Commission (EU), and scholars in technology studies who assess advancements in technology development, commercialization, and transition pathways (Verma and Ramirez-Marquez 2006; European Commission 2014; Kobos et al. 2018).

Scaling Readiness builds on the key principles of Agricultural Innovation Systems (AIS) thinking (Spielman et al. 2009; Hall and Clark 2010; Hounkonnou et al. 2012). Innovation systems are the interlinked set of people, processes, assets, and social institutions that enable (or constrain) the development and scaling of new technologies, products, practices, services, and solutions to deliver impact. A key lesson from conducting innovation systems research in the AR4D sector was that complexity-aware approaches (such as AIS) need to be operationalized or translated into simple tools that can be overseen and managed by program and project teams to guide their practice (Schut et al. 2015). Without such operationalization, AIS approaches can easily be perceived as fuzzy with the risk of people abandoning their willingness to engage with systems approaches altogether.

One of the aims of Scaling Readiness is to support a complexity-aware decision-making process that assists R4D interventions in designing, implementing, and monitoring scaling strategies in a structured and evidence-based way. To this end, Scaling Readiness proposes an iterative cycle of five steps that builds on key principles and concepts that will be further discussed and illustrated in this chapter (Fig. 3.1).

This book chapter has four main objectives:

1. To briefly introduce the key principles and concepts of Scaling Readiness.
2. To explain how Scaling Readiness was used in the CGIAR Research Program on Roots, Tubers and Bananas (RTB) Scaling Fund.
3. To present a case study that illustrates how Scaling Readiness can support the development of better-informed scaling strategies for R4D interventions.
4. To present lessons and recommendations for the further development and use of Scaling Readiness.

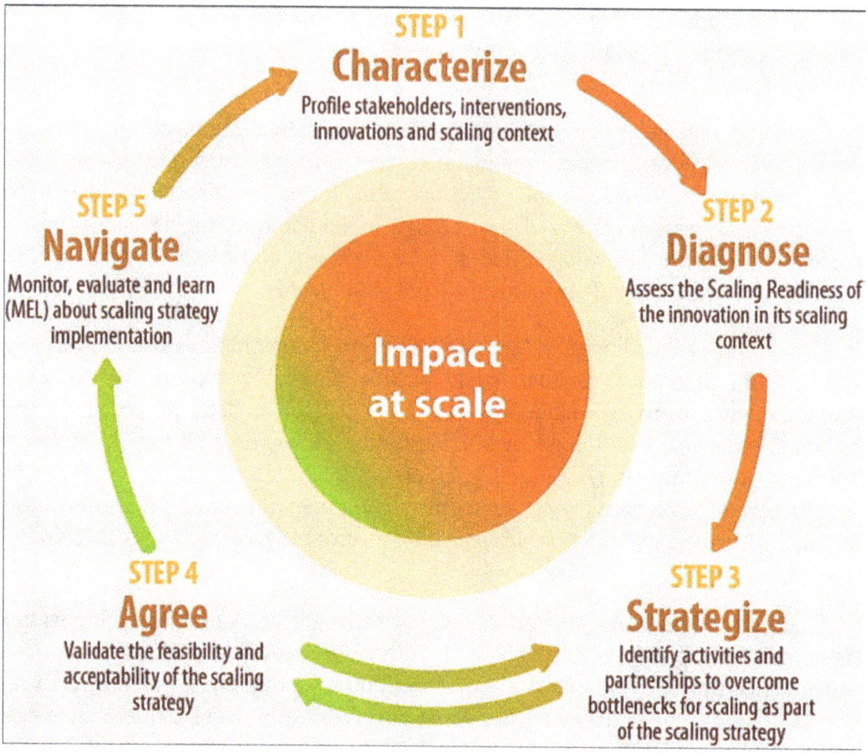

Fig. 3.1 Scaling Readiness proposes a stepwise approach to operationalize AIS thinking in support of the development, implementation, and monitoring of better-informed scaling strategies (Sartas et al. 2020b)

3.2 Scaling Readiness in the CGIAR Research Program on Roots, Tubers and Bananas Scaling Fund

In an attempt to close the gap between the *science* and the *practice* of scaling innovations, RTB developed an institutional innovation to support the scaling of RTB innovations: the *Scaling Fund*. Scaling Readiness was used in two distinct ways in the Scaling Fund: (1) to identify and select scaling-ready RTB innovations and (2) to nurture and support the design, implementation, and monitoring of strategies to scale those RTB innovations. Both will be explained in more detail in the below sections.

It is worthwhile mentioning that the RTB Scaling Fund provided an opportunity for Scaling Readiness not only to develop scaling strategies but also to test and improve the various tools, processes, and workstreams that Scaling Readiness offers.

3.2.1 *Identifying and Selecting Scaling-Ready RTB Innovations*

In 2017, a first call for Scaling Fund project proposals was announced and elicited 12 submissions that were assessed by an independent panel. Five of the proposals were selected to submit full proposals. The evaluation was based on the following assessment criteria:

1. Define and provide *evidence on the scaling readiness* of the selected RTB innovation (referred to as the *core innovation*).
2. Define *site-specific complementary innovations* or enabling conditions that are needed to scale the *core innovation.*
3. Scaling strategy in the proposal is *congruent with existing projects and public and private partners' initiatives.*

From the initial batch of five proposals, the three with the highest scores were awarded a total investment of approximately USD two million to further improve and implement their scaling strategies with their partners. In 2018 and 2019, five additional Scaling Fund projects were funded (Table 3.1).

Table 3.1 Overview of the eight Scaling Fund projects awarded and implemented between 2017 and 2021

RTB Scaling Fund batch	RTB Scaling Fund projects		
2018–2019	1. Single diseased stem removal (SDSR) for BXW banana disease in Burundi, eastern DR Congo, Rwanda, and Uganda	2. Triple S storage process for conserving sweetpotato roots to produce planting material in Ethiopia and Ghana	3. A technology for turning cassava peels into an ingredient of animal feed in Nigeria
2019–2020	1. Orange-fleshed sweetpotato (OFSP) puree for safe and nutritious food products and economic opportunities for women and youths in Kenya, Uganda, and Malawi	2. Approach for flash drying of cassava starch and flour at small scale in Nigeria, DR Congo, and Colombia	3. Rooted apical cuttings in Kenya
2020–2021	1. RTB crop variety validation and diffusion using farmer citizen science in Ghana and Rwanda (TRICOT)	2. A digital fertilizer recommendation service (AKILIMO) in Nigeria, Tanzania, and Rwanda	

3.2.2 *RTB Scaling Strategy Design, Implementation, and Monitoring*

After the initial selection of the Scaling Fund projects, a kick-off and capacity development workshop was organized for each batch of projects (Fig. 3.2). For 2–3 days, the Scaling Fund project teams were trained in the basics of Scaling Readiness and discussed how these would be applied and useful to their own Scaling Fund projects. At the end of the workshop, each of the project teams had a road map for the implementation of their Scaling Fund projects.

To ensure sufficient capacity within the projects to manage scaling processes and implement Scaling Readiness, a key requirement was to assign several people with designated functions, including scaling champions and Scaling Readiness monitors. The scaling champion was primarily responsible for the implementation of scaling strategies and stakeholder engagement plans. They were usually people with a good understanding of the innovation and the local partnership dynamics. Their role was to broker and network for key partners to work together and make scaling happen. The Scaling Readiness monitors were mainly responsible for applying the Scaling Readiness tools to collect and analyze data with the objective to influence decision-making and strategy development with the broader scaling project team. They were usually research assistants with good data collection and analysis skills.

Two different scaling consultants were recruited to backstop the scaling champions and Scaling Readiness monitors and to ensure cross-project learning. During

Fig. 3.2 Group photo of participants representing the three 2019–2020 Scaling Fund project teams during the kick-off workshop in Nairobi, Kenya

the two years of implementation, the scaling champions, Scaling Readiness monitors, and scaling consultant worked closely together with the Scaling Fund project teams.

One of the 2019–2020 Scaling Fund projects (a scaling approach for flash drying of cassava starch and flour at small scale in Nigeria, DR Congo, and Colombia) is used as a case study in this book chapter to explain how Scaling Readiness was used. The cassava flash drying project was selected because of its systematic use and documentation of the Scaling Readiness approach.

3.2.3 *Introduction to the Scaling Fund Cassava Flash Dryer Case Study*

Cassava is a starchy root crop that is a major staple food for people in developing countries. It is grown in tropical regions of the world because of its ability to withstand difficult growing conditions. Cassava in sub-Saharan Africa is generally a subsistence crop, but there is increased commercial interest here in processing cassava flour and for starch production. Rapid perishability of roots requiring agile and efficient processing is one of the greatest challenges facing smallholder cassava farmers and small-scale cassava processors. The most common practice is sun-drying cassava roots to make flour, which is challenging during the extended rainy seasons in the tropics and affects the overall quality of the starch. As a result, farmers and small-scale processors face difficulties to offer their produce to industries that need regular, all-year-round supply and consistent quality for flour and starch production (IITA 2016).

Flash drying, compared to sun drying, enables substantial gains in product quality and productivity by reducing the drying time from between 10 and 48 hours to a few seconds and providing constant drying conditions. Flash dryers are used mostly by large-scale processors (production capacity of >50 tons of starch/day) in countries such as Brazil and Thailand, which have highly developed commercial starch production. Small-scale flash dryers (production capacity of between 1–3 tons of flour/day) are not widely used due to a combination of factors including high energy consumption and production costs. Since 2013, more reliable methods to design energy-efficient flash dryers, based on numerical modelling, have been developed and successfully tested in small-scale pilot flash dryers (Fig. 3.3) that have proven to achieve the same energy efficiency as large-scale industrial flash dryers.

The Scaling Fund Cassava Flash Drying project focusses on three countries where the scaling of small-scale flash dryers has potential: Colombia, DR Congo, and Nigeria. In Nigeria, between 2006 and 2016, prior to the Scaling Fund project, 157 cassava processors had invested in first-generation small-scale flash dryers to produce cassava flour. However, 50% of these are not in use anymore because of the low energy efficiency and subsequent high costs of operation. In the DR Congo, the long rainy season (1300–1900 mm/year) led to strong demands for cost-effective drying solutions such as flash drying. In Colombia, labor-intensive sun drying is costly, motivating cassava processors to seek other drying solutions to increase their

Fig. 3.3 Example of a small-scale cassava flash drying system in Nigeria

production capacity and reduce costs. In each of these countries, partners who were willing to co-invest were identified and brought on board. An initial scaling strategy was proposed in 2019 to train scaling partners (equipment manufacturers and cassava processors) on theoretical and practical aspects of building and operating energy-efficient small-scale flash dryers. In addition, the Scaling Fund project would provide technical support to enable scaling partners to upgrade their existing flash dryers or to invest in new ones. During the project, the Scaling Readiness approach was used to identify bottlenecks and adjust the scaling strategy.

3.3 Principles, Concepts, and Case Study Application of Scaling Readiness

This section introduces the main Scaling Readiness principles and concepts and describes how these were applied in the cassava flash dryer case study following the five Scaling Readiness steps (Fig. 3.1).

3.3.1 Scaling Readiness Step 1: Characterize

During step 1, the project team characterizes the innovation, innovation package, and scaling contexts to explore interdependencies related to the scaling ambitions and aspired impacts.

3.3.1.1 Scaling Readiness Step 1: Principles and Concepts

Scaling Innovation Requires Context-Specific Approaches

A key starting point for Scaling Readiness is that scaling is contextual. Whether something goes to scale and supports the achievement of desired outcomes or impacts depends, for example, on the specific institutional setting (including cultural values, market arrangements, legal frameworks, and policy conditions), on agroecological conditions, and on the interactions that take place within and between networks of interdependent actors and stakeholders (Klerkx et al. 2010; Schut et al. 2015). This implies that an innovation may be scalable in one context but not in another and that scaling strategies successful in one situation may not be effective elsewhere or at another point in time (Baur et al. 2003; Sartas et al. 2019). Similarly, the outcomes of scaling may vary across contexts.

Innovations Never Scale in Isolation

There has been a tendency in both theory and practice to focus on the scaling of a particular – often technological – innovation (Rogers 2003). However, research has shown that the scaling of one particular innovation (e.g., a hybrid seed variety) depends on the simultaneous uptake or enhancement of other practices and services (e.g., seed multiplication, input provision, reorganization of labor, pro-poor credit models, etc.) and/or the downscaling of preexisting practices (e.g., use of open pollinated seed). All of these require attention for successful scaling.

In Scaling Readiness, we consider all innovations or changes that need to take place, including products, technologies, services, and institutional arrangements, and distinguish between "core" and "complementary" innovations. The *core innovation* refers to the initial innovation that an R4D intervention or project aims to develop or scale in order to achieve an assumed societal benefit, for example, the cassava flash dryer. *Complementary innovations* are additional advances or changes in technology, capacity, or policy on which the scaling of the core innovation depends. Together these are labelled the "innovation package."[1] In view of the contextual nature of scaling, the composition of a viable and meaningful innovation package is likely to differ across contexts. That is, the package of core and complementary innovations that is advocated needs to be tailored to different contexts, different target beneficiaries (e.g., specific gender or age groups), and may also need to change over time in view of changing conditions.

On target beneficiaries specifically, there is evidence that different groups in society may face diverse challenges and opportunities in having awareness of, having access to, being able to use, and/or benefitting from innovations. If, for example, market information is provided through a mobile phone-based SMS

[1] See also Bundling innovations to transform agri-food systems. Nature Sustainability 3(12): 973

service, this may benefit men who typically have more control over the household's mobile phone than women. Additional innovations that provide the same market information through a different channel (e.g., printed information provided at a community health center where women regularly visit) may be considered while trying to address some of the underlying inequalities with regard to control over communication assets. This shows that different combinations of core and complementary innovations as part of an innovation package need to be considered for achieving a specific objective or outcome. By being explicit about target beneficiaries during step 1, different types of innovation packages can be considered and designed for different groups of beneficiaries.

3.3.1.2 Scaling Readiness Step 1 in the Cassava Flash Dryer Case Study

During step 1 of Scaling Readiness, the project team characterized the innovation and its context and formulated context-specific innovation package(s) for the three countries. In the flash dryer case study, all core and complementary innovations were defined by the project team and its partners with backstopping from the scaling consultant. A total of 15 core and complementary innovations (Table 3.2) were characterized. The innovations were classified under different innovation types including technologies (i.e., the subcomponent of the machinery), products, services, and institutional arrangements. Since the contexts are different, complementary innovations necessary to scale the flash dryer also differed among the countries.

3.3.2 Scaling Readiness Step 2: Diagnose

During step 2, the project team assesses the current readiness and use of the various core and complementary innovations in the innovation package with the aim of identifying the main bottlenecks toward scaling.

3.3.2.1 Scaling Readiness Step 2 Principles and Concepts

The Scaling Readiness of an Innovation Is a Function of Innovation Readiness and Innovation Use

The notion of "innovation readiness" refers to the demonstrated capacity of an innovation to fulfill its promise or contribute to specific development outcomes. The level of innovation readiness increases as innovations progress from an untested idea to something that has been validated to work in an artificial setting (e.g., a laboratory or controlled project environment) all the way to settings where the innovation has fully matured and has been proven to work under uncontrolled

Table 3.2 Description of the flash dryer core and complementary innovations in the country-specific innovation packages

Core/ complementary innovation	Innovation type	Description	Geographical relevance		
			Nigeria	DR Congo	Colombia
Core innovation	Technology	The **Efficient Flash Dryer** is a pneumatic-conveying dryer that reduces processing cost due to innovative design.	x	x	x
Core innovation sub-component	Technology	Innovation in **the starch/ flour feed system** to the dryer that improves the homogeneity (particle size) of the supply	x	x	x
Core innovation sub-component	Technology	Technological proposal of **mechanical pre-treatment of the raw material** to reach +/- 35% humidity prior to flash drying. Options considered are press and/or centrifuge to remove water	x	x	x
Core innovation sub-component	Technology	**Hot air generator designs** adapted to the requirements and particular conditions of each country to optimize energy consumption and production costs. Depends on type of fuel available (e.g., diesel, gas, agricultural residues/biomass, etc.) and type of burner technology and heat exchanger technology available	x	x	x
Core innovation sub-component	Technology	**New fans/blowers** to achieve sufficient air velocity and flow rate, which improves production capacity	x	x	
Core innovation sub-component	Technology	Adaptation of **drying technology** to the production of sour cassava starch by testing the expansion quality of flash dried sour starch compared to sun-dried			x
Complementary innovation	Product	Preparing a **business plan template** for estimation of costs of investment and operations; estimation of revenues generated; business plans and testing them with manufacturers and processors	x	x	x
Complementary innovation	Product	A spreadsheet **template for assessing availability of raw material and energy** at acceptable cost in the target locations for a cassava starch/flour factory	x	x	x
Complementary innovation	Service	Stimulation of the cassava flour market and **promotion** through social networks; creation of new linkages between actors in the cassava flour value chain and exploration of new domestic and international markets	x		x
Complementary innovation	Service	Information from banks about the conditions and **support** to provide to access investments **loans**	x		
Complementary innovation	Service	**Capacity building** on installing and operating flash dryers	x	x	x
Complementary innovation	Service	Technical forum through facilitated **WhatsApp group** (English) and **technical support** through visits to construction sites	x	x	x
Complementary innovation	Institutional arrangement	A big physical gathering, **forum**, to bring together and promote multi-stakeholder dialogue between entrepreneurs, processors, eqpt manufacturers, funders, government agencies, etc.	x	x	x
Complementary innovation	Institutional arrangement	Assess the feasibility of establishing **cooperatives** of cassava producers (possible support by central bank loans) to ensure sufficient supply of cassava roots to the proposed starch or flour factories	x	x	x
Complementary innovation	Institutional arrangement	**Contracts** between processors, equipment manufacturers, and project teams in order to define the responsibilities, commitments (financial and otherwise), and expected benefits of all parties	x	x	x

Technologies are presented in green, products in blue, services in yellow, and institutional arrangements in orange

conditions (Table 3.3). In contrast to the notion of "technology readiness" that is used by NASA and EU, we use the term "innovation readiness" to signal that the framework can also be applied to measure the maturity of non-technological innovations.

Table 3.3 Innovation readiness and innovation use levels, short names, and basic descriptions of each term (Sartas et al. 2020a)

	Innovation readiness		Innovation use	
Level	Short name	Basic description	Short name	Basic description
0	Idea	Genesis of the innovation. Formulating an idea that an innovation can meet specific goal	None	Innovation is not used for achieving the objective of the intervention in the specific spatial-temporal context where the innovation is to contribute to achieving impact
1	Hypothesis	Conceptual validation of the idea that an innovation can meet specific goals and development of a hypothesis about the initial idea	Intervention team	Innovation is only used by the intervention team who is developing the R4D intervention
2	Basic model (unproven)	Researching the hypothesis that the innovation can meet specific goals using existing basic science evidence	Effective partners (rare)	Innovation has some use by effective partners who are involved in the R4D intervention
3	Basic model (proven)	Validation of principles that the innovation can meet specific goals using existing basic science evidence	Effective partners (common)	Innovation is commonly used by effective partners who are involved in the R4D intervention
4	Application model (unproven)	Researching the capacity of the innovation to meet specific goals using existing applied science evidence	Innovation network (rare)	Innovation has some use by stakeholders who are not directly involved in the R4D intervention but are connected to the effective partners
5	Application model (proven)	Validation of the capacity of the innovation to meet specific goals using existing applied science evidence	Innovation network (common)	Innovation is commonly used by stakeholders who are not directly involved in the R4D intervention but are connected to the effective partners
6	Application (unproven)	Testing of the capacity of the innovation to meet specific goals within a controlled environment that reflects the specific spatial-temporal context in which the innovation is to contribute to achieving impact	Innovation system (rare)	Innovation has some use by stakeholders who work on developing similar, complementary, or competing innovations but who are not directly connected to the effective partners

(continued)

Table 3.3 (continued)

	Innovation readiness		Innovation use	
Level	Short name	Basic description	Short name	Basic description
7	Application (proven)	Validation of the capacity of the innovation to meet specific goals within a controlled environment that reflects the specific spatial-temporal context in which the innovation is to contribute to achieving impact	Innovation system (common)	Innovation is commonly used by stakeholders who are developing similar, complementary, or competing innovations but who are not directly connected to the effective partners
8	Incubation	Testing the capacity of the innovation to meet specific goals or impact in natural/real/uncontrolled conditions in the specific spatial-temporal context in which the innovation is to contribute to achieving impact with support from an R4D	Livelihood system (rare)	Innovation has some use by stakeholders who are not in any way involved in or linked to the development of the R4D innovation
9	Ready	Validation of the capacity of the innovation to meet specific goals or impact in natural/real/uncontrolled conditions in the specific spatial-temporal context in which the innovation is to contribute to achieving impact without support from an R4D	Livelihood system (common)	Innovation is commonly used by stakeholders who are not in any way involved in or linked to the development of the R4D innovation

However, the maturity of an innovation along the innovation readiness scale is not the only factor that is important for understanding and assessing the scalability of an innovation or innovation package in a specific context. There are many examples of innovations with a high level of readiness that were never used at scale. Similarly, there are also examples of innovations that go to scale even if their performance is limited, contested, or poorly documented. Scaling Readiness, therefore, assumes that scalability also depends on the networks in which innovations are embedded and through which their use is supported and advocated (Geels and Schot 2007; Leeuwis and Aarts 2011; Hermans et al. 2017). For example, it makes a difference whether an innovation is only being used by directly incentivized R4D project partners or whether there are partners or beneficiaries that use or promote the innovation independently from the R4D intervention. To capture the degree to which an innovation has penetrated networks, we have introduced the notion of "innovation use" (Table 3.3). The concept also measures the relative magnitude of use (e.g., rare vs. common) to indicate both the scalability potential and actual innovation use at scale. Innovation use is measured using a network analysis approach (Sartas et al. 2018; Sartas et al. 2020a). Scaling readiness, then, must be seen as the function of innovation readiness and innovation use.

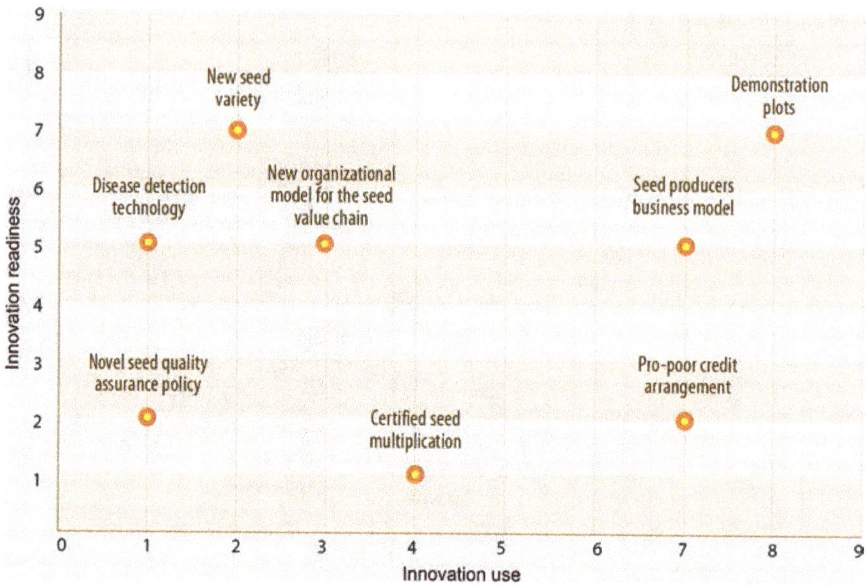

Fig. 3.4 Stylized example of an innovation package (with 8 innovations) that have been assessed for their innovation readiness (y-axis) and innovation use (x-axis) specific to space, time, and goals (Sartas et al. 2020a)

In step 2 of Scaling Readiness, these three concepts are used for diagnostic purposes. With the help of survey techniques, each innovation in a package is assessed for its innovation readiness and innovation use, and evidence of the proclaimed assessment is provided. The scaling readiness of a particular core or complementary innovation results can be found by multiplying the two scores. If the innovation readiness of a particular innovation in the package is at level 3 and innovation use at level 2, the scaling readiness for that innovation is 6 (Fig. 3.4).

3.3.2.2 Scaling Readiness Step 2 in the Cassava Flash Dryer Case Study

Based on the innovation packages defined for the flash dryer scaling work in Colombia, DR Congo, and Nigeria, the project team assessed the innovation readiness and innovation use. The first step to determine the innovation readiness and innovation use level was to collect background information via a short desktop study. The Scaling Readiness monitors gathered information about the available evidence on the readiness and use of the innovation package core and complementary innovations from academic and technical databases and repositories. To complement the desktop study, the project team also collected new data. For the innovations categorized as *core innovation subcomponents*, small-scale processors of cassava flour and starch and flash dryer manufacturers from Colombia, DR Congo, and Nigeria were visited at each location to collect technical information

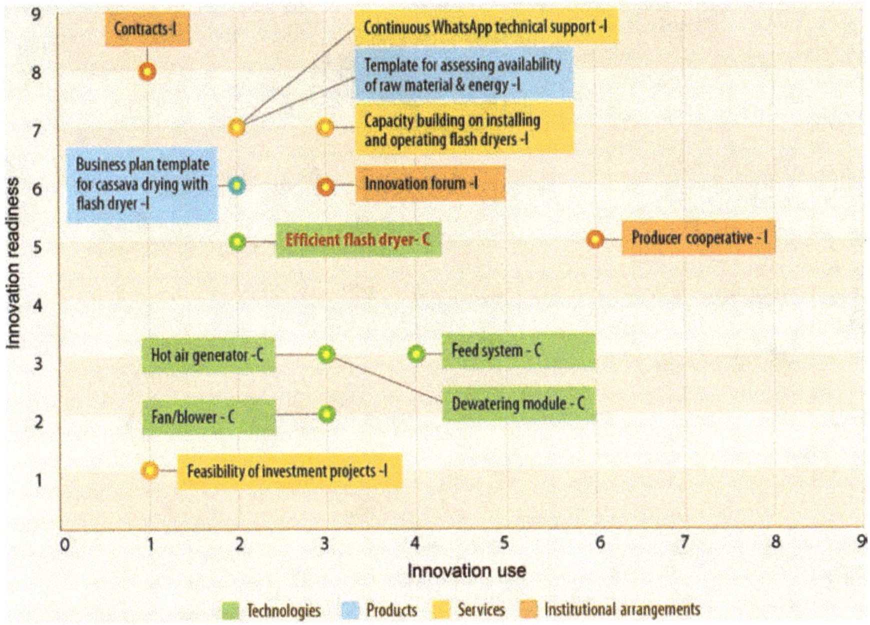

Fig. 3.5 Assessment of the innovation readiness and innovation use of the cassava flash dryer innovation package in Nigeria. (The boxes in Figs. 3.5, 3.6, and 3.7 that include "C" are the core innovation subcomponents of the flash dryer, and "I" refers to complementary innovations. Technologies are presented in green, products in blue, services in yellow, and institutional arrangements in orange.)

used in the analysis. For innovations categorized as *services* and *institutional arrangements* (Table 3.2), information was collected through surveys with different value chain actors, such as bankers, cassava producers, processors, and representatives of government organizations, among others. These surveys were administered during stakeholder meetings and forums in each location.

The information collected via the desktop review, field measurements, and survey results were processed by the Scaling Readiness monitor to determine the innovation readiness and innovation use level for each of the innovations in the innovation package for the three country contexts. A Microsoft Excel template was used to plot the Scaling Readiness graph. The template enabled selection of innovation readiness and innovation use levels from a drop-down list and automatically generated the graph (see Figs. 3.5, 3.6, and 3.7). To validate the results, the Scaling Readiness monitor presented the template to the project team and collaborators and to other key project partners who made their contributions.

The innovation readiness levels ranged between 1 and 8, while the use scores ranged from 1 to 7. These indicated that innovation packages included some new ideas (e.g., organizing an innovation forum in Colombia). Some of the subcomponents of the flash dryer were at the conceptual model stage and not yet validated by the existing applied literature (e.g., fan/blower in Nigeria and DR Congo), while other

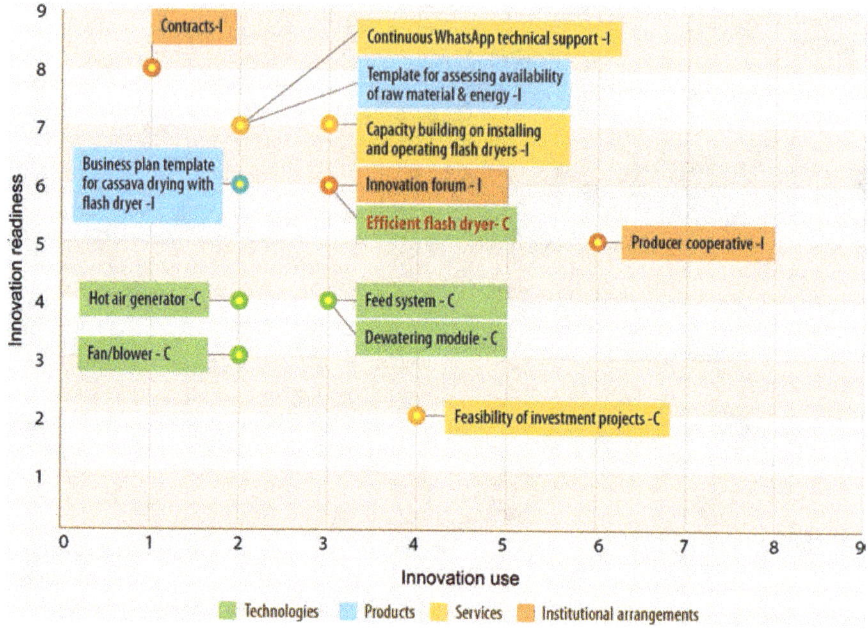

Fig. 3.6 Assessment of the innovation readiness and innovation use of the cassava flash dryer innovation package in DR Congo

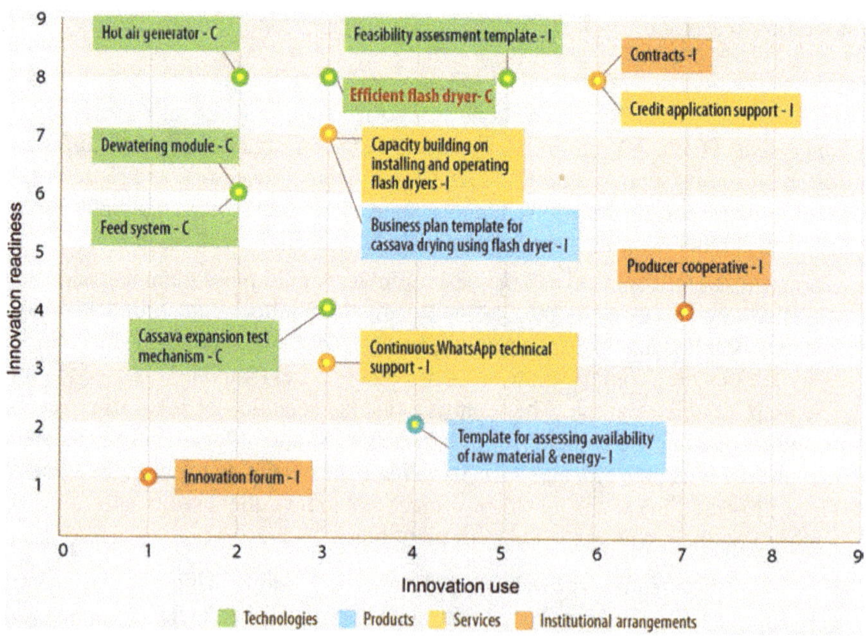

Fig. 3.7 Assessment of the innovation readiness and innovation use of the cassava flash dryer innovation package in Colombia

innovations (e.g., continuous WhatsApp technical support mechanism) were shown to work without a support from R4D interventions in real conditions. Also, the readiness and use levels of some of the components varied among different locations. For example, contracts between the cassava producers and processors were not used in DR Congo and Nigeria beyond those contracts of the project partners, while contracts are commonly used in Colombia.

The diagnosis of the innovation packages indicated that "strengthening the *feasibility of investment projects* through market promotion campaigns for small-scale cassava flour" was the key bottleneck in Nigeria. There was no campaign design that could guide the flash dryer marketing (innovation readiness level 1), and the idea of having flash dryer market promotion campaigns was still under development by the project team (innovation use level 1). In addition, some technical subcomponents of the flash dryer (i.e., hot air generator, dewatering module, and fan/blowers) were assessed at the lower readiness and use, thus needing a strategy for improvement.

In Colombia, the lack of an "innovation forum" was identified as the key bottleneck (Fig. 3.7). "Continuous WhatsApp technical support" for cassava flash dryer installation and use and a "cassava expansion testing" mechanism for the flash dryer were prioritized as the other bottlenecks.

To identify the key partners to overcome the bottlenecks, the flash dryer team used social network analysis. A survey was administered to potential partners, and results were used to characterize stakeholders and partners. Due to the limitations caused by COVID-19, only a few identified partners could be feasibly reached, and those reached did not occupy the most strategic positions in the network. Findings were captured in the stakeholder engagement reports written for DR Congo and Nigeria (Taborda et al. 2020a, 2020b).

3.3.3 Scaling Readiness Step 3: Strategize

During step 3, the project team considers different options and strategies that may be used to address the main bottlenecks to scaling for each innovation package.

3.3.3.1 Scaling Readiness Step 3 Principles and Concepts

Bottlenecks for Scaling Can Be Identified by Assessing Innovation Readiness and Innovation Use

When the core and complementary innovations have been assessed for their level of innovation readiness and innovation use, it becomes pertinent to think about strategies to enhance the readiness of the package as a whole. Scaling Readiness directs most attention to the innovations in the package with the lowest levels of readiness and use, labelled "bottleneck innovations," as they are most likely to limit the scaling of the innovation package. Unless bottlenecks have been addressed, the

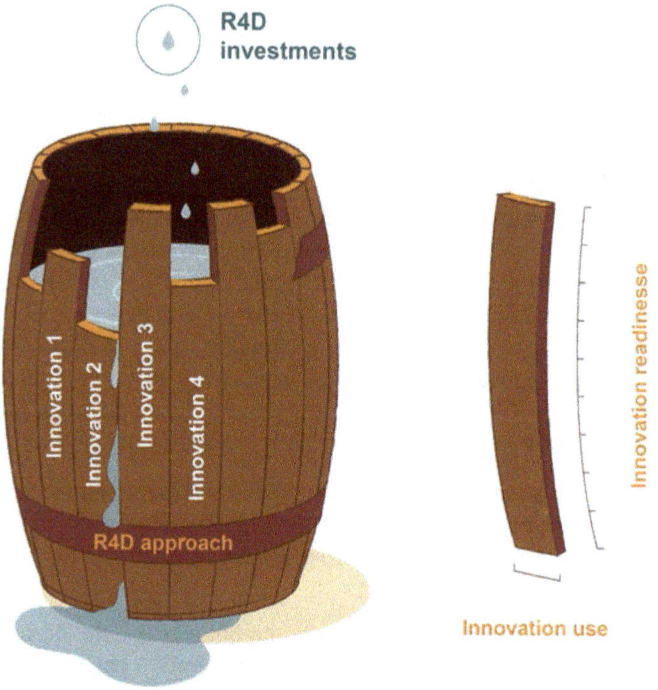

Fig. 3.8 Scaling Readiness barrel to illustrate how innovation(s) with the lowest readiness limits an innovation package's capacity to achieve impact at scale. (Adapted from "von Liebig's barrel," after Whitson and Walster 1912; published in Sartas et al. 2020a, 2020b)

value added to the effort in core or complementary innovations that already have a relatively high innovation readiness and innovation use is low. This point is illustrated in Fig. 3.8, where one can observe that R4D investments (symbolized as water drops) are wasted as they leak away from the lowest stave in the barrel, which symbolizes the bottleneck in the innovation package.

Bottleneck Innovations Can Be Overcome Through Different Strategic Options

Scaling Readiness distinguishes strategic options (i.e., innovation management options) that may be used that address a bottleneck. The choice of an appropriate strategy may be informed by available time, financial and human resources, and organizational mandates and capacities, considering what is feasible and resource efficient (derived from Sartas et al. 2020a):

1. *Substitute:* Can the bottleneck be replaced by another innovation with higher readiness and/or use in the given context?
2. *Outsource*: Are there any organizations or external experts that can more efficiently improve the Scaling Readiness of the bottleneck?

3. *Develop*: Can the intervention team improve the readiness and/or the use by investing available intervention capacities and resources?
4. *Relocate*: Can the intervention objectives be realized more effectively if the intervention is implemented in another location where innovations have higher readiness and use levels?
5. *Reorient*: Can the objective or outcome of the intervention be reconsidered if addressing the bottleneck is not possible and relocation is not an option?
6. *Postpone*: Can scaling the innovation package be achieved at a later point in time?
7. *Stop*: If none of the above strategic options are feasible, should the team consider stopping the intervention?

The strategic options are ranked according to their resource intensity, starting with the least demanding option. The options effectively imply reconsideration of the innovation package and/or the objectives and context of scaling. While we realize that existing project frameworks, budget allocations, and partnership configurations may pose limits to choosing the most sensible and efficient option, Scaling Readiness assumes that considering all options enhances discussion and critical reflection in project teams and thus contributes to the prioritization of relevant and feasible strategies to overcome bottlenecks for scaling. Clearly, the options chosen have further practical implications in terms of who are relevant partners to work with.

3.3.3.2 Scaling Readiness Step 3 in the Cassava Flash Dryer Case Study

From June 2019 onward, the flash dryer project team, scaling champions, and Scaling Readiness monitors explored strategic options for each country. Each bottleneck was discussed, and the most viable options were explored in consultation with key stakeholders and experts. Further information on the strategies is provided in Sect. 3.3.4.2.

3.3.4 Scaling Readiness Step 4: Agree

During step 4, the proposed scaling strategies are shared and discussed with relevant stakeholders to work toward effective collaboration with partners relevant to scaling.

3.3.4.1 Scaling Readiness Step 4 Principles and Concepts

Implementing Scaling Strategies Requires Multi-stakeholder Agreement and Coalition Building

The scaling of an innovation package inherently requires the involvement and cooperation of the various stakeholders. Depending on the package and bottlenecks, these may include policy makers, value chain parties, farmer organizations, community leaders, and/or service (e.g., extension and credit) providers. While AR4D projects depend on the collaboration of such partners to realize their scaling ambition, these parties may not necessarily agree with the proposed scaling strategy, nor may they be ready to take effective action (Sahay and Walsham 2006; Wigboldus et al. 2016). Thus, it is important that initiatives are taken to align interdependent actors and work toward agreement and accommodation on, for example, objectives, strategies, task division, timelines, and investment of resources to enable scaling.

In essence, the process of aligning stakeholders amounts to building an effective coalition that supports change in a particular direction, even if the rationales and interests of stakeholders may only partially overlap (Biggs and Smith 1998; Aarts and Leeuwis 2010). Reaching the necessary degree of accommodation and consensus is far from automatic and often requires active facilitation of learning and negotiation (Leeuwis and Aarts 2011).

3.3.4.2 Scaling Readiness Step 4 in the Flash Dryer Case Study

From June 2019 onward, the flash dryer project team began engaging partners and broader stakeholders in multiple countries to discuss the proposed strategies to improve the scaling readiness of the flash dryer innovation package. In August 2019, combined with a training workshop on small-scale flash drying, the cassava flour processors and equipment manufacturers partners were presented the strategies, and their feedback was collected.

Based on the consultations with partners and key experts and feedback from the workshop, a final strategy for scaling the flash dryer was formulated for the three countries (Table 3.4). To enhance the commitment of the partners to the new strategy, partners were requested to provide their consent and support in writing clearly specifying their intention to participate in the implementation of the scaling action plans. This took the form of an umbrella *participation agreement* explaining the roles, responsibilities, and commitments of the partners (processors, equipment manufacturers) and of the project team to accomplish the goals of the project.

Table 3.4 Scaling Readiness strategic options selected to overcome the bottlenecks for scaling the cassava flash dryer innovation in the different project countries

Bottlenecks (see Figs. 3.5, 3.6, and 3.7)	Bottleneck description	Location	Strategic option	Strategy description
Strengthening the feasibility of investment projects through market promotion campaigns for small-scale cassava flour	Initially, the campaign was conceived to address the lack of markets for small-scale flour producers. However, *the agree step* showed that the dominant majority of the processors did not think increasing the markets is viable in the short term due to the huge efficiency gap and lack of the implementation of the existing local production incentives	Nigeria	Reorient	Since yam prices are higher and yam dryer has higher profitability, the team decided to explore options to use flash dryer for yam
			Postpone	To capitalize a possible favorable change in cassava markets, a demonstration flash dryer was built for promotion and training at the R&D institution FIIRO (although an impact was not anticipated until the end of the project)
Fan/blower, hot air generator, and dewatering subcomponent (only Nigeria)	The capacity of fans to achieve adequate air velocity was too low, resulting in low production capacity. Since the flash dryer team had advanced engineering capabilities, the team chose to optimize the fan/blower designs, share those designs with the manufacturers, and help processors to install them	DR Congo/ Nigeria	Develop	The team has worked on developing the fan and calibrating this tool for efficient drying. Improved fans/blowers were developed and installed (see Figs. 3.5 and 3.6)
Innovation forum	The flash dryer team initially strategized that an innovation forum could increase the awareness of the cassava processors and match them with manufacturers of flash dryers, creating business opportunities	Colombia	Develop and outsource	The team started preparations for the forum and engaged with several organizations to co-organize with. Several activities were planned for outsourcing with other organizations

(continued)

Table 3.4 (continued)

Bottlenecks (see Figs. 3.5, 3.6, and 3.7)	Bottleneck description	Location	Strategic option	Strategy description
Continuous WhatsApp technical support	At the strategizing step, based on the cassava project experience in Africa, the flash dryer team identified that a WhatsApp group could be an instrument for building capacity with the manufacturers and processors	Colombia/ DR Congo/ Nigeria	Develop	The team has opened the WhatsApp group and invited the manufacturers and processors from Africa and Latin America who attended the august workshop in Cali. The information traffic in this network and number of members increased continuously
Cassava expansion mechanism	In Colombia, cassava starch is used for special breads, which requires the expansion. The team has strategized that it can further develop the flash dryer	Colombia	Develop	The team conducted an experiment to measure the effect of various expansion options

3.3.5 Scaling Readiness Step 5: Navigate

During step 5, the project teams monitor the unfolding dynamics in relation to the implementation of agreed-upon scaling strategies and scaling action plans and signal whether major changes in the innovation package configuration or scaling context require a new cycle of Scaling Readiness assessment.

3.3.5.1 Scaling Readiness Step 5 Principles and Concepts

Scaling Projects Need Capacity to Adjust to Emergent Dynamics

When implementing scaling strategies, partners and project teams are likely to meet with unforeseen developments and unintended effects (Hall and Clark 2010; Paina and Peters 2012). This is because scaling contexts are ever-changing and, therefore, can never be fully anticipated (Schot and Geels 2008). It is quite conceivable, for example, that scaling partners meet with new constraints and challenges in their efforts to enhance the Scaling Readiness of a package, or that successful scaling appears to have unwanted side effects for the environment or for specific segments in farming communities. Thus, AR4D interventions require mechanisms to capture and navigate such emergent dynamics. Thus, project teams need to continue to invest in learning and critical reflection when scaling strategies and action plans are implemented.

Relatedly, Scaling Readiness distinguishes between short- and long-term learning cycles and feedback loops (Sartas et al. 2020b). In short-term learning and feedback loops, the focus is on monitoring how the agreed-upon action plans for addressing the bottlenecks are being implemented and on whether plans must be adapted. The long-term learning and feedback loop actually involves a new round of going through the Scaling Readiness cycle, starting with reiterating the Characterize Step 1 and Diagnose Step 2 (See Fig. 3.1). Here, the emphasis is on assessing whether the scaling context has changed and on whether the implementation of scaling strategies has yielded the desired effects. Insights derived from such assessments may result in a reconfiguration of the innovation package, the identification of new bottleneck innovations, and subsequent adaptation of agreed-upon scaling strategies.

3.3.5.2 Scaling Readiness Step 5 in the Cassava Flash Dryer Case Study

The flash dryer team has implemented the agreed-upon strategies presented in Table 3.4 and closely monitored the activities and whether they resulted in the desired improvements in the project. However, travel restrictions due to the COVID-19 pandemic and related closure of businesses necessitated changes in the strategies.

Short-Term Learning and Feedback Loops

In Colombia, the innovation forum was initially postponed (and later cancelled altogether) when the project team realized that the COVID-19 travel restrictions would last much longer than expected. Furthermore, the partners that would co-invest in the flash dryer suspended their commitments to the project and their investment plans. In addition, the development efforts of the flash dryer by the team for the cassava expansion mechanism did not result in desired improvements. As a result, the flash dryer team has decided to stop activities in Colombia and revisited the Scaling Readiness strategic options, adopting a dual strategy:

1. *Relocate and outsource*: Initial consultations with processors in the Dominican Republic showed that there is large interest for the flash dryer; thus, the team decided to relocate there. Since the Dominican Republic was not one of the initial project countries and because the organizational partners of RTB do not have implementation capabilities, the team also decided to outsource the work there to a company called Angavil. The project provided technical support to Angavil to develop its investment plans in flash drying technology for production of cassava flour in the Dominican Republic.

2. *Reorient the flash dryer toward production of high-grade cassava starch* for bioplastics and support an initiative led by the Universidad del Cauca to develop a start-up company in Colombia, funded by Colciencias, the national agency for scientific development. The flash dryer project team decided to provide technical support to this initiative.

In Nigeria, the main bottlenecks were market options for cassava flash drying and inefficient fans/blowers that increased the cost of drying (Fig. 3.5). The customers of cassava flour (e.g., millers, brewers) required much higher volumes than could be supplied by small-scale producers. The buyers typically demand 30 or 60 tons per order, whereas some of the small-scale cassava flour factories can produce up to 1–2 tons per day. This situation led to underutilization of the flash drying capacity since some of the flour producers were too small to be economically viable suppliers for the large-scale buyers: Only 32 flash dryer businesses (out of 64 known flash dryers) were viable users for the cassava flash drying. The fans/blowers of the drying system presented another bottleneck in Nigeria. The team has developed the fans/blowers of the drying system by designing improvements and testing them in the flash drying producers' workshops (Table 3.4).

In the DR Congo, like Nigeria, the main bottleneck was the fans/blowers (Fig. 3.6).

The capacity of fans to achieve adequate air velocity was too low, resulting in low production capacity. Equipment manufacturers acknowledged that they did not have enough experience to build larger fans (due to balancing issues with the rotor) and that they did not know the methods to determine the efficiency of the fan (e.g., air velocity measurements). To address this need, the complementary innovation of flash dryer fans/blowers was added to the innovation package. The team has worked on developing the fan and is calibrating it for efficient drying (Table 3.4).

In both Nigeria and DR Congo, the *heat exchanger* was developed with the goal of driving specific modifications to existing (diesel) heat exchanger designs as well as promoting the manufacture of a new, more efficient heat exchanger design. Out of eight initial private sector scaling partners, three had adopted this innovation by the end of 2019 and increased their processing capacity by 23–50% and profitability by 8–10% – which corresponds to an extra USD 10,000 per year per processor. Cassava producers also benefit from a higher processing capacity, which increases the demand for cassava roots and, hence, economic opportunities for farmers in the regions around cassava factories. Since the most commonly used fuel in Nigeria and DR Congo is diesel, the partners were recommended to change their heating systems to liquid propane gas (LPG), as long as the price of LPG stays competitive in the region. This is a more cost-effective solution (because there is no need to manufacture a heat exchanger) and approximately 10% more efficient with respect to the use of diesel. In 2020, two partners in DR Congo invested in this innovation.

In the Dominican Republic, a scaling action plan was agreed to in early 2020 between the cassava processors, equipment manufacturers and R4D team. The plan was then revised several times through short-term feedback loops. The main feedback was related to the definition of responsibilities for the investment risk, as the cassava processor wanted a guaranteed return on investment, while the equipment manufacturer or the R4D team could not take responsibility for this guarantee due to the novelty of the innovation. Each of the partners reviewed their expectations until the scaling action plan was revised and agreed upon.

Long-Term Learning and Feedback Loops

The short-term learning and changes in the strategies were complemented by an annual assessment of activities. The teams initiated a second Scaling Readiness characterization Step 1, but due to the COVID-19 pandemic, the collection of necessary qualitative information was difficult and this step was suspended.

More generally speaking, in all countries, the COVID-19 pandemic affected most of the programmed activities. For example, two project partners in Latin America (from Colombia and the Dominican Republic) postponed their planned investment in the flash dryer due to the economic impact of the pandemic and a 17% increase in the price of the flash dryer due to the depreciation of the local currency.

The pandemic also caused new challenges and/or bottlenecks to emerge. For example, providing remote technical support for dryer manufacturing, installation, and testing was slower and more complicated than doing it on-site. Some instructions and recommendations provided by video conference were misinterpreted, leading to a need to repeat the work and consequent extra costs. Additionally, in some places in Nigeria and DR Congo, Internet service and electricity supply are patchy, which hindered effective communication with project partners. One strategy to provide efficient remote technical support during the pandemic was the development of protocols and video tutorials for the project partners (e.g., step-by-step assessment methodologies for drying efficiency).

In Nigeria, the implementation of scaling action plans was slower than expected as most project partners delayed their decisions to invest in the flash dryer innovations. Consultation with the different actors in the high-quality cassava flour value chain revealed an emerging bottleneck that had not been identified during the initial Scaling Readiness characterization and diagnosis steps, namely, the rising cost of cassava roots in 2020 that reduced the profit margin of processing. A detailed economic analysis revealed the limited use of high-yielding varieties and good agronomic practices as two of the underlying causes for the limitations.

Although the flash dryer Scaling Fund project did not have the time nor resources to contribute directly to the improvement of cassava production yields scenario, exploring *synergies with other projects* (see Fig. 3.5) became an immediate priority, and a cooperation with the African Development Bank-funded Technologies for African Agricultural Transformation (TAAT) program was established. This program has pursued an objective to provide technical assistance for efficient cassava root production in several African countries, including Nigeria. Overcoming this bottleneck will be key to reducing the cost of cassava roots and related bottlenecks on investment in processing of high-quality cassava flour.

Toward the end of the project, due to the persistence of the COVID-19-related limitations, the team decided to create a virtual platform to scale the flash dryer. Having realized that returning to previous offline work is unlikely, digital solutions were considered one of the best options for advancing the scaling work.

3.4 Reflection on the Use of Scaling Readiness in the RTB Scaling Fund

3.4.1 Reflections on the Use of Scaling Readiness by the Flash Dryer Case Study Team

One of the strengths of the Scaling Readiness approach is that innovation packages are formulated and diagnosed for different scaling contexts. This was acknowledged by the flash dryer project team and was generally appreciated by scaling project teams who used Scaling Readiness. Since each country has its particular context and related bottlenecks, strategies must be adapted to each context to define the most appropriate way to achieve scaling. For example, the teams appreciated analysis required to identify the most appropriate heat generation system for each country to reduce fuel consumption and contribute to the energy efficiency of cassava drying.

To determine the degree of maturity or readiness of the innovations and the level of innovation use, it is necessary to have a deep knowledge of the context. In this regard, one of the lessons learned from the flash dryer case was that it is not enough to carry out this analysis only at the beginning of the project, but that periodic diagnoses must also be carried out, since the context is dynamic and changing, and *emerging* bottlenecks can arise.

Collecting information to design strategies and monitor scaling progress was a challenge in the case study project. Scaling Readiness collects this information through electronic surveys, but it was found that most of the project partners were unresponsive. Some were very busy or did not believe they had the capacity to complete an electronic survey. The challenge will be to develop mechanisms that capture as much information as possible while being user-friendly for the project partners. Collecting information during project meetings or as part of workshops seemed to be more promising in terms of response rates and data quality.

In the course of project implementation, differences emerged between countries in terms of the distribution of responsibilities between processors, equipment manufacturers, and the project team. This dynamic also was felt at the level of co-investment realized and the distribution of financial risks. In DR Congo, cassava processors were confident in the market for cassava flour, which is a staple food product in this country, and, therefore, were more willing to invest in flash dryers and take the financial risks without having a formal agreement. In Nigeria, the market for cassava flour is not functioning smoothly due to a combination of factors, such as low availability of cassava roots at competitive prices for flour production, high processing costs due to low energy efficiency of current flash dryers, and a mismatch between production capacity of cassava processors and demand from large buyers. Consequently, cassava processors were less confident to bear investment risks, with the majority preferring to wait for successful implementation of the flash drying innovation before investing themselves. In Latin America, investment costs (and financial risk) were significantly higher due to higher labor costs and other constraints. In addition, the market for cassava flour is not mature yet. Consequently, cassava processors were not willing to fully take on the investment risks and required that equipment manufacturers, or the flash dryer Scaling Project, offer guarantees against construction cost overruns and potential financial underperformance of the flash dryer system. This led to negotiations and written agreements in the form of a sales contract between the cassava processor and the equipment manufacturer.

These examples underscore the idea that scaling projects always entail financial risk-taking, considering that innovations, by nature, are not yet fully proven with guaranteed return on investment. Therefore, a key bottleneck is finding agreement between project partners who will take on responsibility for these risks. One option to manage this is to identify and select early in the project private partners who are in a position to accept the risks. That is, partners with financial capacity for investment, confidence in the benefits of the innovation, and access to technical expertise to remedy emerging challenges before and after the construction and delivery of the equipment.

3.4.2 *Reflections on the Use of Scaling Readiness in the RTB Scaling Fund*

We offer four main reflections:

1. When the first batch of Scaling Fund projects was selected and approved – early 2018 – the Scaling Readiness approach was still under development. Although the basic principles and concepts of Scaling Readiness were defined, tested, and validated, there were no clear guidelines and workflows that supported its application with partners in controlled conditions – the Scaling Fund projects. Those guidelines and capacity development materials were developed in parallel to Scaling Fund project implementation, which sometimes resulted in confusion (e.g., What is an innovation? What is an innovation package? How to measure and document innovation readiness and innovation use? How to deal with gender and diversity among beneficiaries?). This lack of development also meant a steep learning curve between the Scaling Readiness team and Scaling Fund project teams. The second and especially third batches of Scaling Fund projects

benefitted from those learnings, resulting in a more organized and tailored application of Scaling Readiness. A very concrete spin-off of such learning is the development of a gender-responsible scaling tool for identifying relevant diversity in relation to scaling ambitions (which is currently being designed and tested for use in combination with Scaling Readiness).

2. Scaling of innovation (increasing innovation use) is very different from processes related to designing, testing, and validating the innovation through basic and applied research (improving innovation readiness). For doing the scaling, different skills and competencies, language, organizational space, and incentive structures are required. Many of these skills are very different from those that scientists obtain during their PhD trajectories and require competencies related to being opportunistic, taking risks, and negotiating with scaling partners. After the first year of Scaling Fund implementation, we decided that projects had to identify dedicated scaling champions and Scaling Readiness monitors to ensure that scaling projects were not treated and organized in the same way as science projects. Having dedicated scaling champions and Scaling Readiness monitors clarified the division of tasks and responsibilities in the RTB scaling projects. In addition, capacity development on innovation and scaling processes was very much appreciated by the Scaling Fund project teams.
3. Having senior staff in charge of scaling project design, implementation, and decision-making was not always compatible with best practices for managing scaling projects. The time and responsibilities of senior staff are often fragmented, meaning they need to juggle to a broad variety of science, management, and leadership demands and expectations. Furthermore, scientists are often not on the ground in the context where scaling is desired. The environments in which the cassava flash dryer and other RTB Scaling Fund projects operate are very dynamic and require ongoing *navigation* and *re-strategizing*, which requires operational knowledge. One of the opportunities we see here is to decentralize management and decision-making in scaling projects so that on-the-ground scaling champions can act in a flexible manner based on the analysis and data provided by Scaling Readiness monitors. We have seen that Scaling Fund projects where such a decentralized model for decision making was applied seemed more successful in capitalizing on emerging opportunities and navigating change.
4. When starting Scaling Fund implementation, it was expected that in the first months of the projects, research and scaling partners would go through a cycle of Scaling Readiness steps and start implementing and monitoring their scaling strategies and action plans. This turned out to be very different in actual practice. First, the process of sense-making and capacity development took much longer than expected. Many of the project teams had very different or unclear ideas about their "innovations" and struggled to think critically about what scaling pathways and mechanisms would be required to actually make their products, services, or tools available to end users. Second, working with co-investing scaling partners was essential but also difficult. Scaling partners – especially when they are co-investing – are very deliberate on whether and how

to engage and need to see added value in investing their time and resources in the partnership. As in the flash dryer case, scaling partners may propose risk-sharing strategies – or pull out altogether – if the investment conditions change. It made us realize that the 2-year Scaling Fund projects were essentially about finding common ground between research and scaling partners and creating the space for negotiation, adaptation, and integration that is needed before the actual scaling can happen.

3.5 An Outlook on the Broader Use of Scaling Readiness

The CGIAR Research Program on Roots, Tubers and Bananas has pioneered support and investment in the development and implementation of scaling strategies and partnerships to catalyze the scaling of its innovations. The consequent investment in the Scaling Fund has been timely. The entire CGIAR (2020) is reorganized around an impact-oriented approach, and *scaling* will figure prominently among the different parts of the organization.

Several of the Scaling Fund and Scaling Readiness principles could be embedded in a new way of doing business in CGIAR.

1. *Keep track of innovation readiness and innovation use.* Tracking these elements can support monitoring, prioritization, and resource allocation. Doing this in an evidence-based and structured way can increase transparency, facilitate decision-making, support resource mobilization, and demonstrate return on investment at both the innovation package and portfolio levels. By portfolio level, we mean the management of a broad number of innovation packages and making decisions on which ones to prioritize.
2. *Combine innovation readiness and innovation use in one framework.* With this idea, international organizations, such as the CGIAR, can better link research and development as part of its mandate. Within such a framework, science and applied research focusses on improving innovation readiness in close collaboration with expected beneficiaries and innovation partners, and – once proven to work – scaling can focus on improving innovation use with scaling partners.
3. *Capitalize on the promise of the Scaling Fund co-investment model where research and scaling partners jointly commit funds and capacities to preparing the innovation.* This recommendation would provide a more level playing field between partners and create a higher likelihood that innovations are adapted to become of real value to scaling partners. During the initial stages of sense-making and finding agreement on what the innovation package looks like and which bottlenecks should be prioritized, a *safe incubation space* – such as provided in the Scaling Fund – serves to reduce risk and incentivize partners to find common ground.

4. *Work to create an ecosystem in which the rules and cultures of scaling are different than what people are accustomed to.* This made us realize that the appropriate use of novel approaches such as Scaling Readiness and working in an impact-oriented manner needs to go hand in hand with organizational culture change, capacity development and new incentive structures that reward project teams to prioritize work on bottlenecks in innovation packages. Similarly, strategic options such as reorientation, postponing, relocating, or stopping an intervention when innovation and scaling bottlenecks cannot be overcome should be encouraged, rather than be labelled as a failure as it can avoid wasting of valuable R4D resources.

5. *Ongoing efforts to make Scaling Readiness more sensitive to gender and social differentiation will need to continue.* It seems promising to explore whether and how innovation packages, and scaling strategies can be tailored to groups that are at risk of being excluded.

Acknowledgments This research was largely undertaken as part of, and funded by, the CGIAR Research Program on Roots, Tubers and Bananas (RTB) and supported by CGIAR Trust Fund contributors. The authors thank the following individuals who provided input and feedback on earlier drafts of this chapter: Tom Remington, Jill Lenne, and Christopher Butler. We greatly appreciate Zandra Vasquez's assistance with references and formatting and Claudio Proietti for his support in effectively managing the RTB Scaling Fund.

References

Aarts N, Leeuwis C (2010) Participation and power: reflections on the role of government in land use planning and rural development. J Agric Educ Ext 16(2):131–145

Baur H, Poulter G, Puccioni M, Castro P, Lutzeyer HJ, Krall S (2003) Impact assessment and evaluation in agricultural research for development. Agric Syst 78(2):329–336

Biggs S, Smith G (1998) Beyond methodologies: coalition-building for participatory technology development. World Dev 26(2):239–248

CGIAR (2020) CGIAR performance and results knowledge Hub. Retrieved 10 May 2021, from https://sites.google.com/cgxchange.org/performance/hom

European Commission (2014) Horizon 2020 work program 2014–2015 general annexes revised. Extract from Part 19 - Commission Decision C(2014)4995. ec.europa.eu. European Commission. Retrieved 11 November 2019

Geels FW, Schot J (2007) Typology of sociotechnical transition pathways. Res Policy 36:399–417

Hall A, Clark N (2010) What do complex adaptive systems look like and what are the implications for innovation policy? J Int Dev 22(3):308–324

Hammond J, Rosenblum N, Breseman D, Gorman L, Manners R, van Wijk MT, Sibomana M, Remans R, Vanlauwe B, Schut M (2020) Towards actionable farm typologies: scaling adoption of agricultural inputs in Rwanda. Agric Syst 183:102857

Hermans F, Sartas M, van Schagen B, van Asten P, Schut M (2017) Social network analysis of multi-stakeholder platforms in agricultural research for development: opportunities and constraints for innovation and scaling. PLoS One 12(2):e0169634

Hounkonnou D, Kossou D, Kuyper TW, Leeuwis C, Nederlof ES, Röling N, Sakyi-Dawson O, Traoré M, Van Huis A (2012) An innovation systems approach to institutional change: smallholder development in West Africa. Agric Syst 108(5):74–83

IITA (2016) Making large scale-cassava drying technologies work for small-scale processors. http://bulletin.iita.org/index.php/2016/04/17/making-large-scale-cassava-drying-technologies-work-for-small-scale-processors/

Klerkx L, Aarts N, Leeuwis C (2010) Adaptive management in agricultural innovation systems: the interactions between innovation networks and their environment. Agric Syst 103(2010):390–400

Kobos PH, Malczynski LA, Walker LTN, Borns DJ, Klise GT (2018) Timing is everything: a technology transition framework for regulatory and market readiness levels. Technol Forecast Soc Chang 137:211–225

Leeuwis C, Aarts N (2011) Rethinking communication in innovation processes: creating space for change in complex systems. J Agric Educ Ext 17(1):21–36

Leeuwis C, Klerkx L, Schut M (2018) Reforming the research policy and impact culture in the CGIAR: integrating science and systemic capacity development. Glob Food Sec 16:17–21

Paina L, Peters DH (2012) Understanding pathways for scaling up health services through the lens of complex adaptive systems. Health Policy Plan 27(5):365–373

Rogers EM (2003) Diffusion of innovations. Free Press, New York

Sahay S, Walsham G (2006) Scaling of health information systems in India: challenges and approaches. Inf Technol Dev 12(3):185–200

Sartas M, Schut M, Hermans F, van Asten P, Leeuwis C (2018) Effects of multi-stakeholder platforms on multi-stakeholder innovation networks: implications for research for development interventions targeting innovations at scale. PLoS One 13(6):e0197993

Sartas M, van Asten P, Schut M, McCampbell M, Awori M, Muchunguzi P, Tenywa M, Namazzi S, Sole AA, Thiele G, Proietti C, Devaux A, Leeuwis C (2019) Factors influencing participation dynamics in research for development interventions with multi-stakeholder platforms: a metric approach to studying stakeholder participation. PLoS One 14(11):e0223044

Sartas M, Schut M, Proietti C, Thiele G, Leeuwis C (2020a) Scaling readiness: science and practice of an approach to enhance the impact of research for development. Agric Syst 183:102874

Sartas M, Schut M, van Schagen B, Velasco C, Thiele G, Proietti C, Leeuwis C (2020b) Scaling readiness: concepts, practices, and implementation, CGIAR Research Program on Roots, Tubers and Bananas (RTB). 2020, 217 pp. Available at www.scalingreadiness.org and www.rtb.cgiar.org

Schot J, Geels FW (2008) Strategic niche management and sustainable innovation journeys: theory, findings, research agenda, and policy. Tech Anal Strat Manag 20:537–554

Schut M, Klerkx L, Rodenburg J, Kayeke J, Raboanarielina C, Hinnou LC, Adegbola PY, van Ast A, Bastiaans L (2015) RAAIS: rapid appraisal of agricultural innovation systems (part I). A diagnostic tool for integrated analysis of complex problems and innovation capacity. Agric Syst 132(2015):1–11

Schut M, van Asten P, Okafor C, Hicintuka C, Mapatano S, Nabahungu NL, Kagabo D, Muchunguzi P, Njukwe E, Dontsop-Nguezet PM, Sartas M, Vanlauwe B (2016) Sustainable intensification of agricultural systems in the central African highlands: the need for institutional innovation. Agric Syst 145:165–176

Schut M, Leeuwis C, Thiele G (2020) Science of Scaling: understanding and guiding the scaling of innovation for societal outcomes. Agric Syst 184(102908):1–10

Spielman DJ, Ekboir J, Davis K (2009) The art and science of innovation systems inquiry: applications to sub-Saharan African agriculture. Technol Soc 31(4):399–405

Taborda LA, Adegbite S, Sartas M (2020a) Stakeholder engagement strategy for Scaling small scale flash drying system for cassava starch and flour production in Nigeria, CIAT and IITA

Taborda LA, Lukombo SS, Sartas M (2020b) Stakeholder engagement strategy for Scaling small scale flash drying system for cassava starch and flour production in Democratic Republic of Congo, CIAT and IITA

Verma D, Ramirez-Marquez J (2006) From TRL to SRL: the concept of systems readiness levels. Presented at the Systems Engineering Research, www.boardmansauser.com

Whitson AR, Walster HL (1912) Soils and soil fertility. St. Paul, MN: Webb. 73 p. OCLC 1593332. 100. Illustration of limiting factors

Wigboldus S, Klerkx L, Leeuwis C, Schut M, Muilerman S, Jochemsen H (2016) Systemic perspectives on scaling agricultural innovations. A review. Agron Sustain Develop 36(3):1–20

Woltering L, Fehlenberg K, Gerard B, Ubels J, Cooley L (2019) Scaling – from "reaching many" to sustainable systems change at scale: a critical shift in mindset. Agric Syst 176:102652

Part II
RT&B Crop Products: Innovations in Processing, Marketing and Distribution

4

Flash-Dryer Innovations: Development and Innovation

Thierry Tran, Adebayo Abass, Luis Alejandro Taborda Andrade, Arnaud Chapuis, Marcelo Precoppe, Laurent Adinsi, Alexandre Bouniol, Makuachukwu Ojide, Suraju Adeyemi Adegbite, Simon Singi Lukombo, Murat Sartas, Béla Teeken, Apollin Fotso Kuate, Robert Ndjouenkeu, Martín Moreno, John Belalcázar, Luis Augusto Becerra López-Lavalle, and Dominique Dufour

Abstract The development and scaling out of flash-dryer innovations for more efficient, small-scale production of high-quality cassava flour (HQCF) and starch is described. The diagnoses of cassava-processing SMEs (small and medium enterprises) revealed their energy expenditures for drying were considerably higher than those of large-scale industrial companies, which was mostly due to suboptimal design of flash-drying systems. As a result, small-scale production of cassava starch and HQCF often incurs high production costs, incompatible with market prices of final products. Taking stock of this situation, RTB scientists have developed several innovations to optimize energy efficiency and costs, including a longer drying pipe, reengineered heat exchanger, larger blower for higher air velocity, and a higher product/air ratio. This was based on numerical modelling to determine the key design features of energy-efficient flash dryers, followed by construction and demonstration of a pilot-scale prototype. As a result, improved small-scale flash dryers are now being scaled out to the private sector in various countries, using the Scaling Readiness framework and achieving 10–15% gains in productivity and incomes. A method for diagnosis of process efficiency is also described, to identify technical bottlenecks and to document and measure the outcomes and impacts during the implementation of scaling-out projects.

T. Tran (✉)
Alliance Bioversity-CIAT, Cali, Colombia

CIRAD UMR Qualisud, Montpellier, France
e-mail: thierry.tran@cirad.fr

A. Abass
International Institute of Tropical Agriculture (IITA), Dar-Es-Salaam, Tanzania
e-mail: a.abass@cgiar.org

4.1 Introduction

Postharvest processing is a key link between cassava crops and end products, determining the quality and food safety for consumers. Postharvest processing also contributes to food security by stabilizing and increasing the shelf life of

L. A. T. Andrade
Centro Internacional de Agricultura Tropical (CIAT), Cali, Colombia
Universidad Nacional de Colombia, Bogotá, Colombia
e-mail: latabordaa@unal.edu.co

A. Chapuis · D. Dufour
CIRAD, Montpellier, France
e-mail: arnaud.chapuis@cirad.fr; dominique.dufour@cirad.fr

M. Precoppe
Natural Resources Institute (NRI), Chatham, UK
e-mail: M.Precoppe@greenwich.ac.uk

L. Adinsi
Université d'Abomey-Calavi, Faculté des Sciences Agronomiques (UAC-FSA), Jericho, Benin

A. Bouniol
Université d'Abomey-Calavi, Faculté des Sciences Agronomiques (UAC-FSA), Jericho, Benin

CIRAD UMR Qualisud, Cotonou, Benin
e-mail: alexandre.bouniol@cirad.fr

M. Ojide
Alex Ekwueme Federal University Ndufu-Alike-Ikwo (FUNAI), Ebonyi state, Nigeria

S. A. Adegbite
Federal Institute of Industrial Research Oshodi (FIIRO), Lagos, Nigeria

S. S. Lukombo
International Institute of Tropical Agriculture (IITA), Kinshasa, DR, Congo
e-mail: S.Lukombo@cgiar.org

M. Sartas
Wageningen University, Wageningen, Netherlands

International Institute of Tropical Agriculture (IITA), Kigali, Rwanda
e-mail: M.Sartas@cgiar.org; murat.sartas@wur.nl

B. Teeken
International Institute of Tropical Agriculture (IITA), Ibadan, Nigeria
e-mail: B.Teeken@cgiar.org

A. F. Kuate
International Institute of Tropical Agriculture (IITA), Yaoundé, Cameroon
e-mail: A.Fotso@cgiar.org

R. Ndjouenkeu
Ecole Nationale Supérieure des Sciences Agro-Industrielles, University of Ngaoundéré, Ngaoundéré, Cameroon

M. Moreno
Universidad del Valle (Univalle), Cali, Colombia
e-mail: martin.moreno@correounivalle.edu.co

J. Belalcázar · L. A. Becerra López-Lavalle
Alliance Bioversity-CIAT, Cali, Colombia
e-mail: l.a.becerra@cgiar.org

perishable crops such as cassava and other roots, tubers, and bananas. Beyond product quality and safety, processing is also crucial for improving the sustainability of cassava value chains: by optimizing processing technologies, it is possible to reuse product waste while reducing energy and water consumption, product losses, production costs, and the overall environmental footprint of cassava industries. Processing is a crucial way for many small and medium farmer-processors to add value to their crops. From a gender perspective, especially in Africa, women are important in processing roles, so improving processing technologies can potentially improve their working conditions and increase their incomes. Women can benefit from processing innovations that match their dominant type of technology, i.e., small and medium scale (Taiwo and Fasoyiro 2015; Teeken et al. 2018, 2021; Ndjouenkeu et al. 2021; Thiele et al. 2021). Processing innovations may also influence gender roles if new technologies change the processing scale or the level of involvement of different stakeholders in the value chain (Forsythe et al. 2015, 2016).

Postharvest processing of cassava involves several steps, called unit operations, to transform fresh roots into several finished products. Some unit operations are common to most cassava products, including washing, peeling, rasping/grating, and dewatering (pressing). Other unit operations are specific to certain products, particularly the cooking or drying operations such as toasting for gari, steaming for attiéké, sun drying for fufu and conventional cassava flour, and flash drying for high-quality cassava flour (HQCF) and starch. Lactic fermentation is also a specific unit operation used to preserve and confer a desirable sour taste to the end products (gari, fufu, chikwangue). Cassava processing generates large amounts of by-products (e.g., peels, fiber bagasse, and wastewater), which typically accumulate around the processing sites or pollute local water systems (Tran et al. 2015). Expansion of postharvest technologies must therefore include strategies for by-product management and processing to reduce environmental impacts and to create additional income streams for processors, most commonly animal feed, but also biomass to energy, such as solid fuel and biogas (Okudoh et al. 2014; Ozoegwu et al. 2017; Patrizi et al. 2020; Yank et al. 2016; Zvinavashe et al. 2011).

Over the past 20 years, research for development projects has identified the high potential of interventions in postharvest processing to improve working conditions and reduce inefficiencies in cassava value chains across Africa, Latin America, and Asia (Abass et al. 2013, 2017, 2018; Awoyale et al. 2017; Kuye et al. 2011; Dufour et al. 2002). Cassava processing is mainly a small-scale operation, in particular in Africa, at factories handling less than 5 tons of fresh roots per day or at household level. Current small-scale technologies are often suboptimal, resulting in high use of resources (energy, water, firewood, product losses) and high production costs (Adenle et al. 2017; Da et al. 2013; Kitinoja et al. 2011; Nzudie et al. 2020). Improving the efficiency of small-scale equipment, their design, safety, and ergonomics is therefore essential for the competitiveness of cassava value chains (Abass et al. 2018; Edeh et al. 2020; Oni and Oyelade 2014; Taiwo 2006; Nweke 1994), together with the ability of small-scale processors to access investment capital (Taborda 2018). The potential impacts of such interventions are to empower small-scale processors and farming communities to process more of their crops close to the production areas, resulting in more affordable products and increased food security and resilience against fluctuations in the prices of imported foods.

Improving cassava processing also encompasses food safety. African food products tend to have high microbiological and chemical contamination levels exceeding regulatory limits. Relative to industrialized countries, various deficiencies at government, sector, retail, and company levels affect performance of food safety management systems. Collective efforts needed to address food safety include developing stringent certification standards and product specifications, improving hygiene and control of raw materials, enhancing monitoring systems, developing quality assurance and supportive administrative structures, developing risk-based legislative frameworks, and strengthening food safety authorities (Kussaga et al. 2014).

As a staple crop, cassava cultivation and processing tend to expand in lockstep with growing populations in developing countries. From this perspective, improving the efficiency of cassava processing technologies is crucial to minimize resource consumption and environmental footprint and to avoid a business-as-usual scenario whereby current inefficient technologies multiply to handle increasing volumes of cassava, leading to unsustainable levels of pollution and resource depletion in the face of climate change and other environmental risks. This issue also has important gender ramifications: "An analysis of distributive impacts of the environment on human well-being cannot ignore features such as gender. [...] Women and girls often carry a disproportionate burden from environmental degradation compared to men" (UNEP 2007, p 15).

In recent years, researchers federated under CGIAR's Research Program on Roots, Tubers and Bananas (RTB) have made vital contributions to a range of products and to the links between processing, product quality, and consumer expectations (Escobar et al. 2021; Adinsi et al. 2019; Alamu et al. 2019; Bouniol et al. 2021; Luna et al. 2021). This chapter presents the reengineering and scaling out of flash-drying technology for small-scale cassava processing, supported by the development of a methodological framework for R&D on postharvest processing of cassava.

4.2 Optimization of Energy Efficiency for Drying of Starch and Flours

In countries with industrial production of cassava starch, such as Thailand, Vietnam, Brazil, and Paraguay, large factories produce 50–400 tons of starch per day. These factories use large, highly efficient pneumatic-conveying dryers known as flash dryers (Brennan 2011; Sriroth et al. 2000; Aichayawanich et al. 2011). Even so, drying often has the highest processing cost due to the energy needed for heating. In contrast, in most cassava-producing countries, less intensive cassava production, limited transport infrastructure, and a tradition of small-scale household or family-based working units predominantly using manual labor do not permit the development of large factories. As a result, most cassava processing into flour and starch for food consumption or industrial applications is done at small scale, with an average production capacity of 1–3 tons/day (Adegbite et al. 2019). Such processing systems are beneficial in providing jobs for many in the community, including women, thus contributing to social stability and reducing migration to cities. From that perspective, proposed technical innovations and related efficiency gains imply a trade-off between the risk of job losses and the prospect of lower production costs and consequently higher incomes for processors and lower food prices for consumers. At any rate, small-scale processing is often affected by low investment capacity and limited repair-and-maintenance services, slowing down the adoption of technical innovations.

For drying, many processors still rely on sun drying, which is low cost but has limitations, including being subject to the weather, the limited availability of flat surfaces, and contamination from animals, dust, or microbes (Precoppe et al. 2020; Alonso et al. 2012). This is a particular disadvantage to access larger markets that need regular, all-year-round supply and consistent quality (Dziedzoave et al. 2006). Attempts to solve these limitations in the past 20 years have involved scaling down the design of large-scale flash dryers. However, so far, there has been little use of the resulting models of locally built, small-scale flash dryers due to high energy consumption related to suboptimal design and consequently high production costs incompatible with market prices of final products. For instance, in Nigeria between 2006 and 2016, 157 cassava processors have invested in "first-generation" flash dryers to produce HQCF, most of which are no longer in use in spite of some improvements (Ojide et al. 2021).

Taking stock of this situation and given the strategic importance of efficient post-harvest processing for the sustainable expansion of cassava value chains, research initiated in 2013 by RTB identified several critical points to optimize energy efficiency and minimize the operating costs of small-scale flash dryers. This research was based on numerical modelling of flash drying (Chapuis et al. 2017) and validated through the construction and testing of a pilot-scale dryer (capacity, 100 kg/h) at the International Center for Tropical Agriculture (CIAT, Colombia) that achieves the same energy efficiency as large-scale industrial flash dryers. The Scaling Readiness framework was then applied to scale out this innovation to cassava processors.

4.3 What Is an Efficient Dryer? Definition and Expression of Drying Efficiency and Affecting Factors

Drying of flour and starch consists of removing water by thermal treatment, inducing a phase change, generally from liquid to gas. Flash dryers use convective drying, i.e., applying a hot air stream to the product. Evaporation of liquid water requires an important quantity of energy, called latent heat of vaporization. Its value of 2500 kJ/

kg of water is thermodynamically incompressible and sets the minimum energy requirement for drying. In real drying systems, energy losses are unavoidable, so an energy consumption of 3000–4000 kJ/kg water is considered efficient.

In practice, energy is delivered to the dryer by heating the air flowing into the system. Gas burners are convenient because they produce combustion gases clean enough to be in contact with the product, even food. Otherwise, heating systems using fuels such as diesel, fuel oil, or biomass require a burner coupled to a heat exchanger. These systems are more common because they allow using cheaper and widely available fuels, thus reducing operating costs.

During drying, the hot air releases heat to evaporate water and cools down as it absorbs water vapor. Efficient drying, i.e., maximum water evaporation, requires *good exchange area and contact time* between the air and the product and *good mixing (turbulence)* to promote heat exchanges. Many drying technologies are available depending on the product to dry. Flash dryers are among the most efficient for granular materials (Crapiste and Rotstein 1997), as they maximize the exchange area between air and suspended solids (Saravacos and Kostaropoulos 2016). The capacity of air to hold water vapor is limited by phase equilibrium and increases with temperature (hot air can hold more water). Therefore, the *higher the initial temperature* of the drying air, the better, without damaging the product. As an illustration, air heated to 200 °C can absorb 60 g water/kg air, three times more than at 80 °C (20 g water/kg air). Under optimum drying conditions, the exhaust air is saturated with moisture, and the product has reached the target moisture for a long shelf life, i.e., 12–13% wet basis (wb) in the case of starch and flour.

Finally, mechanical dewatering to reduce the water content before drying is also important, as it requires much less energy and is more cost-effective than drying by evaporation (Mujumdar 2006). As an illustration, drying HQCF to 12% moisture content from an initial 40% (after good dewatering) requires 1860 kJ/kg of final product, assuming a dryer using 3500 kJ/kg of water evaporated. If starting from an initial 50% moisture content (after inefficient dewatering), drying requires markedly more energy at 3020 kJ/kg of product.

4.4 Key Design Components for Efficient Small-Scale Flash Drying

4.4.1 Surveys on Flash-Dryer Designs and Energy Efficiency

The first action of the RTB initiative to improve energy efficiency of small-scale flash drying consisted of a survey of processors owning flash dryers in various countries (Thailand, Vietnam, Tanzania, Nigeria, Colombia, and Paraguay) to identify common design features underpinning energy efficiency, such as dimensions and operating conditions (Tran et al. 2015a, b, c; Saengchan et al. 2015). The physical properties of cassava particles before and after flash drying were also studied

(Romuli et al. 2017). While the drying principle and key components remained the same across all flash dryers (Fig. 4.1), important variations in designs, dimensions, and operating conditions were observed (Fig. 4.2, Table 4.1). Large-scale dryers had longer drying pipes (20–57 meters) than small-scale ones (7–15 m). The quantity of air used for drying varied more than tenfold between dryers, with air-to-product ratio ranging from 6 to 75 kg of air/kg of product (Table 4.1). Small-scale dryers tended to use more hot air per kg of product, resulting in higher energy use and drying costs.

Standardized drying costs ranged from 23 USD up to 400 USD/t of product (Table 4.1), assuming for comparison purpose that all the flash dryers used diesel fuel (energy density, 40 MJ/L; price, 1 USD/L). In reality, flash dryers are powered by various fuels, renewable or nonrenewable, including wood, biogas, agricultural biomass (palm oil kernel, cashew nut kernels), liquefied petroleum gas (LPG), heavy fuel oil, diesel, black oil (a mixture of kerosene and used motor oil), and coal. Processors owning flash dryers with high fuel consumption manage to keep drying operations profitable by using cheaper fuels than diesel. Nevertheless, flash dryers with high fuel consumption reduce the profitability of processing and contribute to small-scale factories abandoning production of HQCF or starch, often at a significant loss on their investment. In Nigeria, a recent survey of 41 HQCF processors to

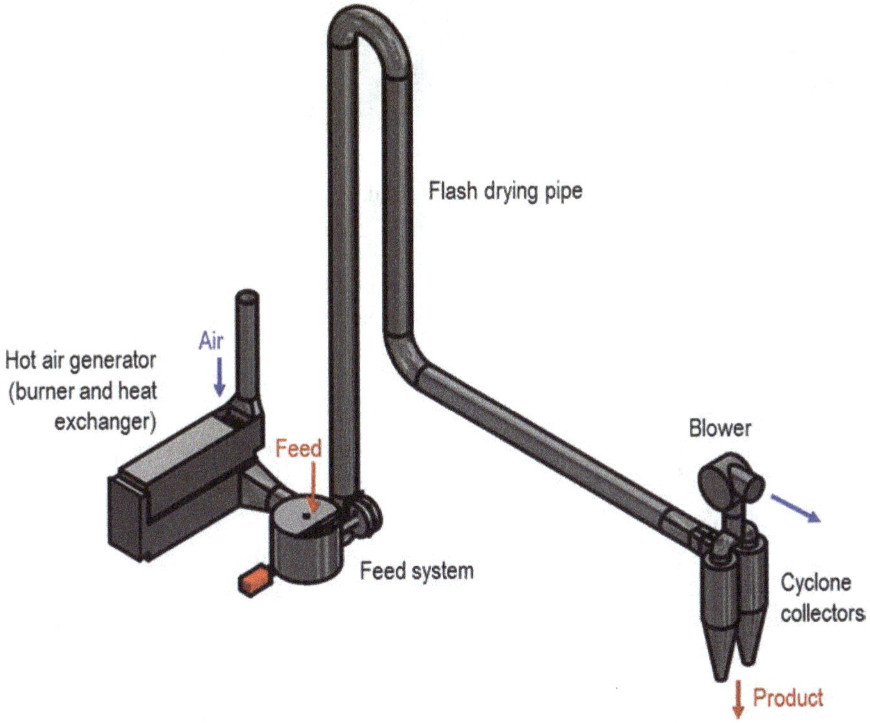

Fig. 4.1 Key components common to all flash-dryer designs

Fig. 4.2 Diversity of flash-dryer designs in Thailand, Nigeria, Paraguay (dryer model from Brazil), Paraguay (dryer model from Sweden), and Argentina

ascertain current perceptions on flash dryers found that 50% of them have stopped producing in the past 10 years and that most see their market prospects as limited (Ojide et al. 2021). Processors who remain in operation have done so through cost-saving measures: being near the source of raw materials and the end-product market, thus cutting transportation costs; replacing petroleum-based fuels with more economical coal or agricultural biomass; and diversification to produce gari or fufu when there is no demand for HQCF.

Given the small profit margins in HQCF production, small-scale processors currently select cheaper fuels, without considering the environmental costs of more polluting fuels. As a result, nonrenewable hydrocarbon fuels are still widely used. Cleaner renewable fuels such as biogas may become viable options for small-scale processing in the future, as long as the greater use of renewable energies does not also incentivize the burning of firewood, which would exacerbate deforestation. Renewable fuels are possible, and investments by large-scale cassava starch factories to generate biogas from factory wastewater can substantially reduce both drying costs and environmental impacts (Hansupalak et al. 2016).

Table 4.1 Diversity of dimensions and operating conditions of flash dryers in selected countries

	Unit	C-III	C-IV	TH-1	TH-2	TH-3	TH-4	N-2	N-3	T-1
Country		Paraguay	Paraguay	Thailand	Thailand	Thailand	Thailand	Nigeria	Nigeria	Tanzania
Capacity	t prod. wb/h	4.57	4.36	7.54	14.9	8.30	11.2	0.04	0.05	0.2
Total pipe length	m	33	20.7	46.0	57.7	45.0	44	12.5	7.7	13.6
Pipe diameter	m	1.2	1.5	1.16	1.50	1.14	1.25	0.38	0.75	0.36
Product input moisture	% wb	34	40	34	36	35	35	39	48	–
Product output moisture	% wb	13	13	13	13	13	13	5	10	–
Air flow	t air/h	69.9	52.2	49.2	114.9	63.8	66.2	2.2	3.8	2.3
Air/product ratio	kg/kg	15.1	12.2	6.4	7.5	7.5	5.8	56.1	75.3	11.6
Input temperature	°C	130	155	173	177	170	200	222	137	275
Output temperature	°C	70	62	52	55	54	59	62	86	
Energy use	MJ/t prod.	1503	1534	933	1109	1077	996	16,087	8212	3055
Energy use[a]	MJ/t water	4613	3541	3080	3087	3182	3035	15,393	11,238	4377
Drying costs[b]	USD/t prod.	37.6	38.4	23.3	27.7	26.9	22.6	402	205	76

[a]Energy used in the drying pipe only. The energy losses at the burner and heat exchanger are not considered

[b]For comparison purposes, the drying costs presented above have been standardized assuming diesel as fuel (40 MJ/L diesel) costing 1 USD/L

The surveys revealed the absence of a consensus design among flash dryers. These variations in design and ultimately in energy efficiency and drying costs suggest that there is a huge potential for optimization, with benefits in terms of production costs and environmental impacts (fossil fuel consumption and greenhouse gas emissions). The surveys were thus followed by an in-depth study of the technical characteristics needed to achieve energy-efficient flash drying.

4.4.2 Numerical Modeling of Flash Drying Provided Design Guidelines for Energy Efficiency

Several physical phenomena govern drying kinetics of a particle in suspension: heat transferred from the air to the particle causes water to evaporate from the surface, which, in turn, causes water within the particle to diffuse toward the surface. At the same time, the particle is entrained by the air flow along the drying pipe. These phenomena are described using conservation equations of mass, momentum, and heat (Mujumdar 2006). In the case of starch and flour, the slowest phenomenon is the diffusion of water within the particles; therefore, the size of the particles and their residence time in the dryer are critical for quick drying.

Applying this theoretical framework to flash drying, a numerical model was developed to simulate various designs and predict energy efficiency and final moisture content of cassava flour and starch (Chapuis et al. 2017). The technical data collected during the surveys (Sect 4.4.1) were used to validate that the model gave accurate drying predictions. Deviations between predicted data and actual data were satisfactorily low (less than 10%) for moisture content of the end product, air temperature after drying, and energy consumption.

The numerical model was applied to investigate optimum dryer designs that minimize energy consumption and maximize the quantity of water evaporated per unit mass of hot air (Sect. 2.1). The main challenge was to identify a configuration where the hot air carried just enough energy to dry the product flow rate (thermodynamics design objective) and the particles stayed in contact with the hot air long enough for complete heat and water transfers (kinetics design objective). Numerical simulations showed that such configuration could be attained through the following key design components:

(i) The target production capacity of the flash dryer defines the flow rate of water to evaporate, or *drying rate* (kg water evaporated/h):
*Drying rate = MassFlowProductIn*MoistureIn - MassFlowProductOut*MoistureOut*, where *MassFlowProductIn = MassFlowProductOut*(1-MoistureOut)/(1-MoistureIn).*
(ii) The drying rate in turn defines the minimum energy input for drying and hence the required quantity of hot air to enter the dryer, or *air mass flow*. For efficient

drying, the minimum energy input (or specific energy use) should aim for 3000 kJ/kg of water evaporated. Then.

*HeatInput (kW) = SpecEnergyUse (kJ/kg)*DryingRate (kg/h)/3600* and

*AirMassFlow (kg/h) = 3600*Heatinput / (HeatCapacity*(TempAirIn-TempAirAmb))*, where HeatCapacity = 1.01 kJ/K/kg.

For instance, drying from 38% to 12% moisture, using initial air temperature of 180 °C, requires about 10 kg of air per kg of final product.

(iii) *Air velocity* must remain high enough to keep the particles in suspension. For particles of typical size distribution (10 μm up to 2 mm), the threshold for suspension is 8–9 m/s. Air velocity should be set above the threshold, typically 12 m/s, as a safety margin. A long drying pipe (20 m or longer) gives flexibility to use higher air velocities to increase the dryer capacity without affecting energy efficiency significantly.

(iv) Air mass flow and air velocity in turn define the *section and diameter of the pipe*:

*PipeSection (m^2) = AirMassFlow/(3600*AirDensity*AirVelocity)*, with AirMassFlow in kg/h and AirVelocity in m/s. AirDensity (kg/m^3) depends on temperature (0.77 kg/m^3 at 180 °C).

(v) The *residence time* for the product to fully dry depends on the particle size, initial moisture content, initial air temperature, and air velocity. For typical drying conditions of cassava starch or HQCF, a residence time of 2 s is recommended.

(vi) The *length of the drying pipe* is defined by residence time and air velocity, but also input moisture content and initial air temperature. A length of 17 meters is a minimum, and 20 meters or longer is recommended for energy efficiency and stability of operations. A dryer with a longer pipe is more versatile for handling variations in product moisture, air temperature, and air velocity while delivering stable moisture content in the final product.

(vii) Finally, initial *air temperature* should be as high as possible without damaging the product (Sect. 2.1), with optimum between 180 and 200 °C.

These findings helped to interpret the surveys, according to which longer pipe and lower air/product ratio achieve better energy efficiency (Table 4.1 and Fig. 4.3). Approaches to improve the less efficient, small-scale flash dryers were identified: First, all small-scale flash dryers surveyed had short drying pipes (7.7–13.6 m), which should be extended to 20 m to increase residence time and water evaporation. Second, with a longer drying pipe, the air/product ratio can be reduced by increasing the feed rate of the product.

To validate these conclusions, the next step was to build and test a prototype of a small-scale flash dryer with adjustable pipe length and adjustable air velocity.

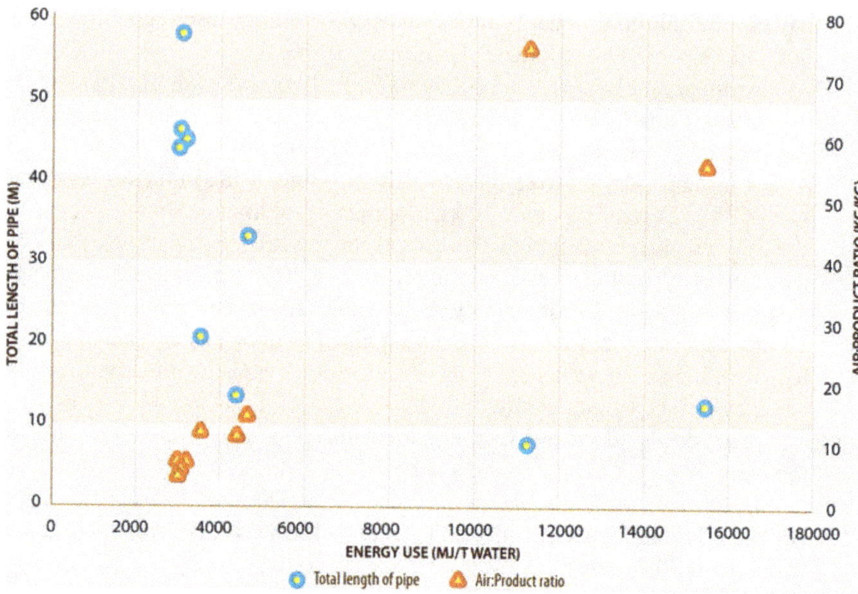

Fig. 4.3 High energy use by flash dryers is correlated with shorter pipe and high air/product ratio. Each point represents a different model of flash dryer (Table 4.1)

4.4.3 *Prototype Flash Dryer Confirmed That High Energy Efficiency at Small Scale Is Achievable*

Numerical simulations indicated that energy-efficient flash drying at small scale was achievable. To validate these results through actual experiments, a prototype flash dryer (100 kg/h capacity) was designed and built at CIAT (Cali, Colombia) in 2017 (Fig. 4.4). In addition to integrating the key design components (Sect. 4.4.2), the following elements were also essential for operating the dryer:

- *Regulated feed system:* Wet cassava flour or starch is a semi-sticky powder with a tendency to form clumps during handling. Accordingly, the feed system was composed of a cylindrical hopper equipped with stirring paddles to prevent caking and powder bridging and discharging into an endless screw conveyor leading to the drying pipe. A pin mill located between the screw and the drying pipe disaggregated clumps, to minimize particle size and maximize the exchange surface. The speed of the endless screw was regulated by a feedback loop maintaining the temperature of the air stable after drying (setpoint 55 °C). This regulation maintained a stable moisture of the end product, maximized energy efficiency, and reduced fuel consumption and costs. This control strategy can be implemented by manually reading the air temperature after drying and by manually adjusting the feeding rate; however, an automatic temperature controller (PID) is more effective.

Fig. 4.4 Flash-dryer prototype installed at CIAT

- *Conveying system and collection system of the dry product:* Inside the drying pipe, air and product are kept in movement with a blower, which can be located at the beginning or at the end (positive or negative pressure, respectively). Negative pressure is recommended because the blower is located after the cyclone that collects the dry product, which requires less power as the air is cooler (55 °C). Moreover, after the cyclone, the air is free of solid particles, which reduces wear and breakdowns and allows using a blower with more efficient curved blades (as opposed to straight blades). On the other hand, negative pressure requires that the bottom output of the cyclone be closed with a rotary airlock valve to prevent the product from being re-entrained into the cyclone.
- In a positive-pressure system, the blower is located at the beginning of the pipe after the feed system and needs to be bigger to move the same quantity of air as an equivalent negative-pressure system due to the higher temperature (150–180 °C) and lower air density. The power consumption of a positive-pressure blower is thus 50–100% higher. Another drawback of positive pressure is that the product passes through the blower, which requires more maintenance and allows only less efficient straight blades to avoid accumulation of product on the blades. On the other hand, in positive-pressure dryers, a single flap valve at the bottom of the cyclone is sufficient instead of a rotary airlock valve.

- *Hot air generator (burner and heat exchanger):* Common fuels such as diesel or biomass generate fumes and particles, which must be kept separated from the product that is being dried. For small-scale dryers, this is achieved through an air-air heat exchanger installed with the burner. Current models are mostly single pass concentric pipe type, with sub-optimum efficiency. The RTB team developed an improved heat exchanger integrating the following:
  - (i) Counter-current flow of the combustion fumes and fresh air
  - (ii) Increased heat exchange surface through a bundle of thin smoke pipes for the combustion fumes, around which fresh air circulates in a series of chambers with chicanes
  - (iii) Increased turbulence of the flow of fresh air forced to pass through the chicanes (Fig. 4.5). This design increased air outlet temperature while reducing fuel consumption

Alternatively, if a distribution network of LPG is available, the heat exchanger system can be replaced with a direct-combustion LPG gas burner, which reduces fuel consumption (by 10–15%) and investment costs compared to a system with a heat exchanger.

The performance of the prototype flash dryer was assessed under diverse operating conditions (Table 4.2). High energy efficiency was achieved with a pipe longer than 17 m, with specific energy consumption in the range 3300–4000 kJ/kg of water evaporated (Table 4.2). These figures are slightly higher than for large-scale dryers due to higher surface-to-volume ratio and consequently higher heat losses in smaller equipment (Kemp 2012). The best energy efficiency was obtained with a 20-m-long pipe (3268 kJ/kg water) and high air velocity (24 m/s).

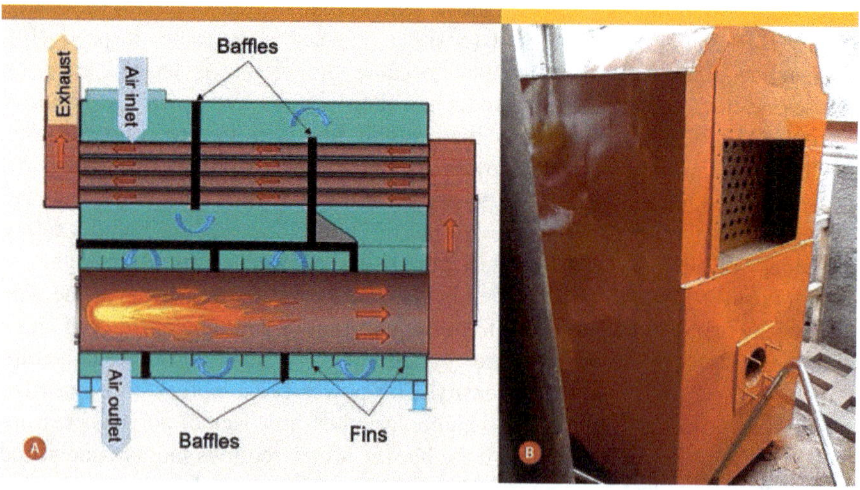

Fig. 4.5 (**a**) Cut view of the counter-current improved air-air heat exchanger developed by RTB; (**b**) one of the six heat exchangers built during the scaling-out phase of the project. (Source: Authors and Agrimac Ltd, DR Congo)

Table 4.2 Performance of the prototype small-scale flash dryer built at CIAT (Cali, Colombia) under various operating conditions (fixed pipe diameter, 16 cm)

		Trial #							
	Unit	1	2	3	4	5	6	7	8
Pipe length	m	17.2	17.2	17.2	17.2	20.3	20.3	27.2	27.2
Inlet air velocity[a]	m/s	11	11	23	23	11	24	22	22
Inlet air temperature	°C	140	180	180	180	140	180	140	140
Outlet air temperature	°C	47.5	50.6	53.6	56.5	47.1	52.6	51.9	52.0
Product initial moisture	% wb	31.6	31.4	36.8	36.9	34.0	38.5	36.7	36.4
Product final moisture	% wb	13.4	13.3	13.5	11.9	14.3	13.6	12.9	13.3
Product feed rate	kg/h	77	99	164	152	67	160	122	124
Specific energy consumption	kJ/kg water	3563	3610	3340	3486	3979	3268	3399	3417

[a]Air velocity is provided at the corresponding inlet air temperature

The numerical model predicted that high air velocity would increase specific energy use due to shorter residence time and consequently incomplete drying. On the contrary, actual performance tests indicated that increasing air velocity is not necessarily detrimental (Table 4.2). A possible explanation is that higher air velocity also reduces particle size (a factor not included in the numerical model) through increased collisions, thus intensifying the drying. Consequently, when the drying pipe is long enough (>20 m), increasing air velocity may substantially increase drying capacity, without affecting energy efficiency.

Design guidelines and tools resulting from this RTB initiative are available from the authors upon request.

4.5 From Prototype to Technology Transfer: Optimization of Feeding Rate for Energy Efficiency of a Small-Scale Flash Dryer in Ghana

4.5.1 Introduction

After demonstrating the energy efficiency of the small-scale flash-dryer prototype, the RTB research team moved to transfer these findings to the private sector. Processors from several countries expressed interest (DR Congo, Nigeria, Ghana, Cameroon, Uganda, Tanzania, Colombia, Brazil, and the Dominican Republic) to reduce operating costs and/or increase production of HQCF or starch. Reasons cited included growing demand for HQCF for industry or to replace sun-dried cassava flour, which consumers increasingly see as lower quality due to contamination during drying. Underpinning these trends was ongoing economic development with the emergence of a larger middle-class and more urban consumers with higher incomes. In 2018, a small-scale energy-efficient flash dryer was developed and put into commercial use in the Central Region of Ghana. This provided an opportunity to further evaluate the effect of feeding rate on energy efficiency, following previous findings by Precoppe et al. (2016).

4.5.2 *Materials and Methods*

Drying procedure and processing equipment At the partner processing center in Ghana (Tropical Starch Company Ltd.), cassava roots are peeled and washed manually and mashed using a mechanical grater. The mash is dewatered with a screw-operated press. The resulting press cake is pulverized into wet grits using another mechanical grater. The wet grits are fed into a pneumatic dryer, and the resulting dried grits are milled into flour. The pneumatic dryer was developed with joint funding from RTB and the Cassava: Adding Value for Africa project (CAVA II). CAVA-II partners (equipment manufacturers and Tropical Starch Company Ltd.) were given training on the design and construction of small-scale pneumatic dryers, including the key design components described in Sect. 4.4.2. The equipment was constructed with stainless steel and thermally insulated with 50-mm-thick mineral wool, shielded with aluminum sheeting (Fig. 4.6b). The drying duct had a diameter of 0.18 m and a length of 24.84 m. To reduce the height of the equipment, the drying duct was divided into seven vertical meandering sections (Fig. 4.6a). Air flow was forced by a centrifugal 7.5-kW blower located at the start of the drying duct, just after the feeding point (positive-pressure system). The design point used for the

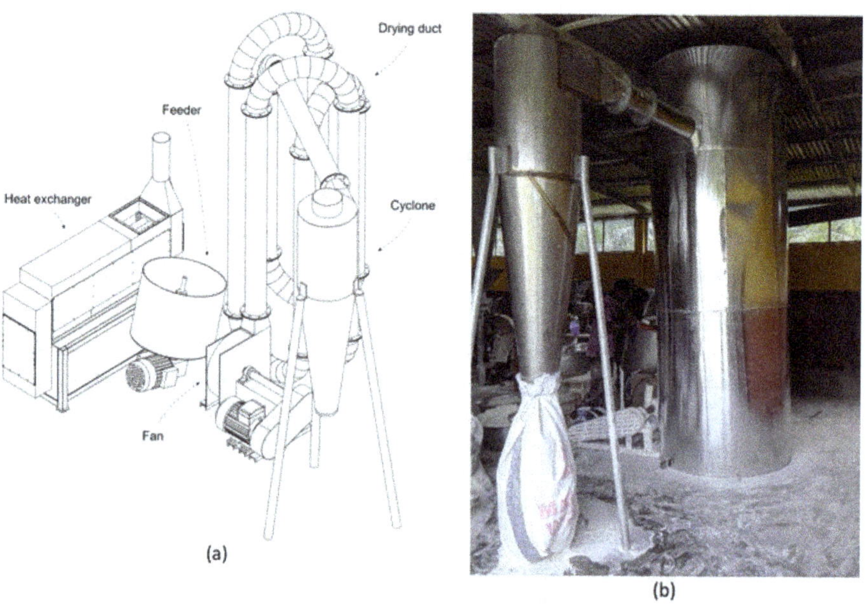

Fig. 4.6 (**a**) Isometric view and (**b**) the pneumatic dryer in Ghana

blower construction was an air mass flow rate (dry basis, db) of 550 kg/h and a static pressure of 10.2 hPa. Air was heated by a 70-kW diesel burner (model B14; Bairan, Wenling, China) and the improved RTB heat exchanger (Fig. 4.5). Drying air temperature was thermostatically controlled with the sensor placed at the dryer inlet, between the heat exchanger and the feeding point.

Experimental design The energy performance of the dryer was first evaluated following the processing center's drying procedure, with no interventions other than sampling material and recording drying conditions. Based on the data collected, the feeding rate was adjusted to optimum value, and the data collection was repeated. Data were collected over six consecutive days (recording 5 hours/day), the first 3 days with the original feeding rate and the subsequent 3 days with the optimum feeding rate. Data collection started after the dryer had been in operation for 1 hour to ensure steady-state conditions.

Statistical analyses Statistical analyses were performed with SAS 9.4 (SAS Institute Inc., Cary, NC, USA) software, following a comparative experimental design (Precoppe et al. 2016) with three replicates and two treatments: before and after the adjustment on the feeding rate. One-way analysis of variance (ANOVA) was performed, and to determine whether the pairwise difference comparisons were significantly different, Fisher's Least Significant Difference (LSD) was used at a 5% level of significance.

Data collection Temperatures, relative humidity, pressures, and air velocities were measured at various points on the dryer (Fig. 4.7) with suitable sensors connected to a wireless data logger (LOG-HC2-RC, Rotronic) recording values at 1-minute intervals. Mass flow rates of wet and dry cassava grits were measured using a digital balance (AWB120; Avery Weigh-Tronix, Smethwick, UK) and a chronometer. Samples of wet and dried cassava grits were collected at 1-hour intervals for moisture content analysis (3 h at 103 ± 1 °C in a convection oven) according to AOAC 935.29 (AOAC 1998).

4.5.3 Calculations

Energy performance Psychrometric calculations used the formulas provided by British Standard (2004). Energy performance was calculated as described by Precoppe et al. (2016). The *solid mass flow rate* ($\dot{m}_{dm}$, db) was calculated based on $\dot{m}_{ws}$ and the moisture content of the wet cassava grit. *Heat input rate to the dryer* (Q_{in}) was calculated based on air temperature, relative humidity, and air flow rate ($\dot{m}_{air}$) at the dryer inlet. *Water evaporation rate* ($\dot{m}_{w}$) was calculated using $\dot{m}_{ws}$ and the difference in moisture content between the wet and dried cassava grits. *Specific heat consumption* (q_s) was calculated dividing Q_{in} by $\dot{m}_{w}$, and energy efficiency (η) was calculated dividing the heat used for moisture evaporation (Q_{w}) by Q_{in}. Finally,

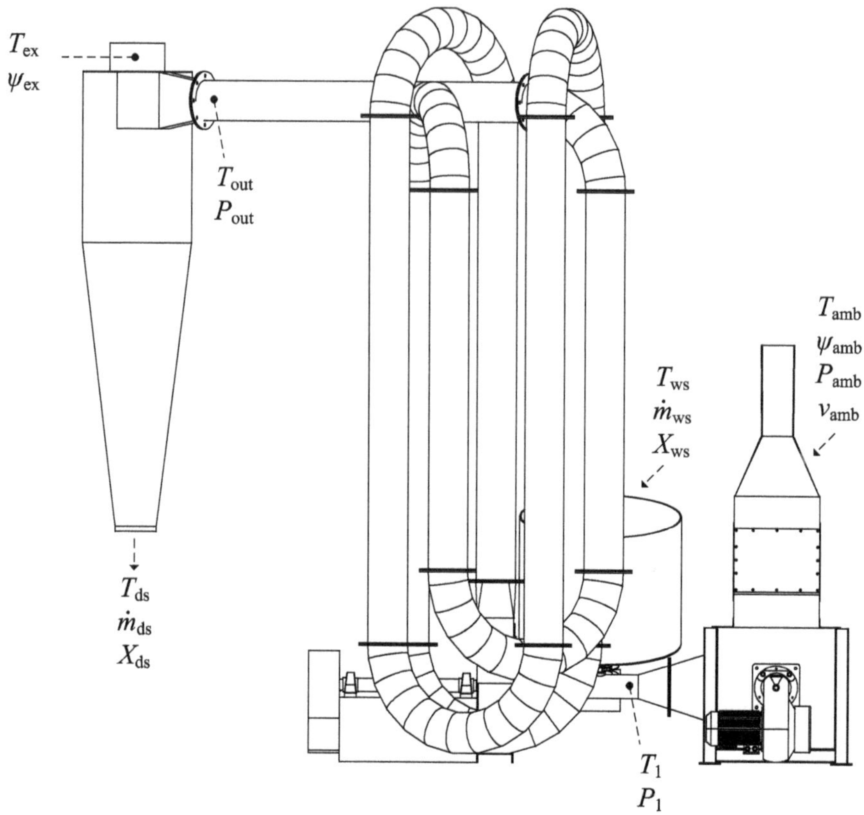

Fig. 4.7 Measurements on the pneumatic dryer in Ghana. *T*, temperature; ψ, relative humidity; *P*, pressure; *v*, air velocity; $\dot{m}$, mass flow rate; *X*, moisture content; ws, wet cassava grits; ds, dry cassava grits; amb, 1, out, and ex refer to air characteristics at various points of the dryer

specific heat utilization (q_u), i.e., energy consumption per kg of dried product, was calculated dividing Q_{in} by $\dot{m}_{dm}$.

Optimum feeding rate The dryer's original feeding rate was 65.7 ± 11.4 kg.h^{-1}. This rate was adopted by trial and error by the operators of the flash dryer over a period of 1 year of use. With this feeding rate, T_{out} was 67.3 ± 4.9 °C, and ψ_{out}, calculated using T_{ex} and ψ_{ex}, was 37.6 ± 8.7%. At this temperature and relative humidity, the enthalpy at the dryer outlet (h_{out}) was 249.9 ± 15.5 kJ.kg^{-1}. Based on these values, the lowest allowable air temperature at the dryer outlet was T^*_{out} = 60 °C (Precoppe et al. 2016). Keeping h_{out} unchanged $\left(h_{out} = h^*_{out}\right)$ and reducing T^*_{out} to 60 °C, the highest allowable relative humidity at the dryer outlet $\left(\psi^*_{out}\right)$ can be raised to $\psi^*_{out} = 53\%$. Keeping the other settings unchanged and monitoring the air at the dryer outlet, the feeding rate was gradually increased until T_{out} approached T^*_{out} and ψ_{out} approached ψ^*_{out}. At this point, the optimum feeding rate was determined to be 98.6 kg.h^{-1}, a 50% increase compared to the original configuration.

4.5.4 *Results and Discussion Regarding the Small-Scale Flash Dryer in Ghana*

Dryer operating conditions Adjusting the dryer to optimum feeding rate increased the output of dried cassava grits from 42.2 ± 7.3 $kg.h^{-1}$ to 65.0 ± 5.5 $kg.h^{-1}$, a 54% increase in productivity without increasing energy consumption (Table 4.3). Input parameters independent from the feeding rate, i.e., air temperature at inlet and moisture content of wet cassava grits, were unchanged between the treatments. The moisture content of the dried cassava grits (0.14 $kg.kg^{-1}$) was unchanged as well, confirming that it is possible to increase the feeding rate to its optimum without altering the quality of the end product.

The T_1 values seemed high for food drying (Kudra 2009); nevertheless, short residence times in pneumatic dryers allow the use of high temperatures without jeopardizing product quality (Pakowski and Mujumdar 2014). When drying cassava, the temperature of the product (T_{ds}) must remain below 56 °C to avoid starch gelatinization (Breuninger et al. 2009), which was realized both before and after adjusting the feeding rate.

Energy performance The water evaporation rate is driven by heat input to the dryer and by the amount of material (Kudra 2009). Increasing the feeding rate to optimum value increased the evaporation rate, which significantly improved the energy performance of the dryer (Table 4.3). Heat input rate is independent from the feeding rate and hence remained unchanged.

Energy efficiency at optimum feeding rate (72.1%) was close to the top of the range reported for pneumatic dryers (50–75%, Strumiłło et al. 2014) but still lower than the values reported by Sriroth et al. (2000) for large-scale pneumatic dryers. Small-scale dryers inevitably have an unfavorable surface-to-volume ratio that results in higher heat losses (Kemp 2012). The higher feeding rate increased the solid loading ratio of the drying air; nevertheless, the conveying mode remained in the dilute phase, and pneumatic transport was not jeopardized.

4.5.5 *Conclusions: The Small-Scale Flash Dryer in Ghana*

This case study in a commercial setting confirmed that small-scale flash dryers for HQCF can achieve high energy efficiency similar to large-scale dryers. Adjusting the feeding rate to its optimum value also improved energy performance significantly and is easier to implement than adjusting the air flow rate as it does not require hardware changes.

Table 4.3 Operating conditions and energy performance of a pneumatic dryer processing cassava in Ghana before and after adjustment to the feeding rate

Adjustment to the feeding rate	Unit	Before	After
Solid mass flow rate ($\dot{m}_{dm}$)	kg.h^{-1}, db	37.3^a ± 6.1	56.0^b ± 5.1
Air temperature at dryer inlet (T_1)	°C	236.6^a ± 5.1	238.4^a ± 9.5
Wet cassava grits moisture content (X_{ws})	kg.kg^{-1}, db	0.76^a ± 0.03	0.75^a ± 0.06
Dried cassava grits moisture content (X_{ds})	kg.kg^{-1}, db	0.14^a ± 0.04	0.14^a ± 0.06
Dried cassava grits temperature (T_{ds})	°C	55.9^a ± 3.3	54.4^a ± 2.2
Air temperature at dryer outlet (T_{out})	°C	67.3^a ± 4.9	60.7^b ± 3.5
Relative humidity at dryer outlet (ψ_{out})	%	37.6^a ± 8.7	55.3^b ± 9.2
Heat input rate	kW	31.4^a ± 1.0	32.4^a ± 5.2
Water evaporation rate	kg.h^{-1}	26.4^a ± 4.4	33.6^b ± 5.6
Specific heat consumption	kJ.kg^{-1} water evaporated	4388^a ± 716	3509^b ± 527
Energy efficiency[a]	%	57.8^a ± 9.0	72.1^b ± 9.9
Specific heat utilization	kJ.kg^{-1} dried product	2746^a ± 437	1798^b ± 278
Solid loading ratio	g.kg^{-1}	70.1^a ± 11.5	105.7^b ± 9.5

In each line, means with different superscript letters are significantly different by Fisher's Least Significant Difference test at 5% level of significance

[a]% of the heat input to the dryer that was used for water evaporation (not including the energy efficiency of the air heating system)

4.6 Applying Scaling Readiness to Scale Out Flash-Drying Innovations to DR Congo and Nigeria: Successes and Lessons Learned

4.6.1 *Scaling Readiness Framework*

Scaling Readiness is a stepwise approach for analyzing the characteristics of innovations from a scaling-out perspective, diagnosing the issues (bottlenecks) that hinder scaling out, developing optimum scaling strategies, building common agreements with the key stakeholders on the strategy, monitoring and learning from the implementation of the agreed strategies, and finally updating the strategies accordingly (Sartas et al. 2020a). Scaling Readiness analyzes the core innovation by breaking it down into complementary innovation components. "Innovation Readiness" refers to the demonstrated capacity of the innovation to fulfill its contribution to development outcomes in specific locations. This is presented in nine stages showing progress from an untested idea to a fully mature, proven innovation. "Innovation Use" indicates the level of use of the innovation or innovation package by the project members, partners, and society. This shows progressively broader levels of use beginning with the intervention team who develops the innovation to its widespread use by users who are completely unconnected with the team or their partners. "Scaling Readiness" of an innovation is a function of innovation readiness and innovation use. Table 4.4 provides summary definitions for each level of readiness and use, adapted from Sartas et al. (2020a).

Table 4.4 Summary definitions of levels of innovation readiness and use (Sartas et al. 2020a)

Stage	Innovation readiness	Innovation use
1	Idea	Intervention team
2	Basic model (testing)	Direct partners (rare)
3	Basic model (proven)	Direct partners (common)
4	Application model (testing)	Secondary partners (rare)
5	Application model (proven)	Secondary partners (common)
6	Application (testing)	Unconnected developers (rare)
7	Application (proven)	Unconnected developers (common)
8	Innovation (testing)	Unconnected users (rare)
9	Innovation (proven)	Unconnected users (common)

Readiness and use levels are evaluated using conceptual, applied, and experimental evidence. Scaling Readiness helps to identify the gaps in the design of the innovations and to prioritize the research and engineering work to address these gaps (Sartas et al. 2020b). The Scaling Readiness framework is discussed in more detail in Chap. 3.

Building on previous steps (Sects. 4.4 and 4.5), the RTB research team applied Scaling Readiness to support the scaling out of the flash-dryer innovation in the private sector in DR Congo and Nigeria, as a first case example of applying Scaling Readiness to a postharvest processing industry (Table 4.5).

4.6.2 *Identification of the Innovation Package and Determination of the Innovation Readiness and Innovation Use of Each Innovation Component*

In the RTB flash-dryer case, the complementary innovations were identified collaboratively with stakeholders. Small-scale processors of cassava flour (1–3 tons of flour/day) and equipment manufacturers from Nigeria and DR Congo were interviewed to generate a baseline of the technical and socioeconomic conditions of potential scaling project partners. In Nigeria, the cassava processors were selected based on being active in the cassava processing business or continuous processing operations in their factories. Equipment manufacturers were pre-selected based on experience in the commercial manufacture and sale of flash dryers. For sustainability of the flash-dryer innovation, the final selection gave priority to relatively younger manufacturers willing to learn new technology innovations. In DR Congo, flash-dryer manufacturing and use are new but with high potential to improve the quality and food safety of conventional sun-dried cassava flour. Only two local equipment manufacturers had exposure to flash dryers, and three processors were using flash dryers. All three processors and one equipment manufacturer were selected. The scaling partners explained the constraints of the current flash-dryer

Table 4.5 Levels of innovation readiness (adapted from Sartas et al. 2020a) applied to flash drying

Stage	Innovation readiness	Flash-dryer example	Evidence proving completion of the level
1	Idea	Flash drying can increase the energy efficiency of drying HQCF and starch	Crapiste and Rotstein (1997); Saravacos and Kostaropoulos (2016)
2	Basic model (testing)	An energy-efficient flash dryer has five key components (hot air generator, feed system, flash-drying pipe, blower, cyclone collector)	Figure 4.1
3	Basic model (proven)	An updated version of the basic model increased the efficiency of cassava flash drying, as validated with numerical simulations covering multiple scenarios	Chapuis et al. (2017); Sect. 4.4
4	Application model (testing)	A prototype based on the proven basic model and numerical model findings was designed	(Available upon request)
5	Application model (proven)	Multiple versions of the prototype designs were assessed by RTB researchers and finalized by technical drawings with specifications	Figure 4.4 3D drawing, Figs. 4.5a and 4.6a
6	Application (testing)	CIAT prototype was built based on the technical drawings in Colombia	Figure 4.4 photos; Figs. 4.5b and 4.6b
7	Application (proven)	CIAT prototype assessment was shown to improve the efficiency of flash drying significantly	Table 4.2
8	Innovation (testing)	Commercial small-scale, energy-efficient flash dryer was built in Ghana by RTB and CAVA II	Section 4.5, Tables 4.3 and 4.4, Fig. 4.7
9	Innovation (proven)	Testing of flash dryers in commercial setups is ongoing	N.A.

models (in particular high fuel consumption) and the expected performance improvements.

Based on on-site visits and interviews with stakeholders of the cassava value chains, scaling partners, and researchers, 16 innovation components were identified, of which ten were essential for the scaling process: four core technical innovations and six complementary innovations related to capacity building and the socioeconomic context of scaling out. The Scaling Readiness assessment showed varying levels of innovation readiness and use depending on the country (Fig. 4.8).

In both Nigeria and DR Congo, the levels of use and readiness of the technical components were relatively low (2–4 and 3–6, respectively). Nigeria had slightly higher levels of use of the feed system, hot air generator, and blower components. On the other hand, the readiness of the core innovation, i.e., efficient flash dryer, was higher in DR Congo. The main reason for this disparity was because at the beginning of the project, Nigeria had markedly more first-generation flash dryers in operation (higher use), albeit less efficient (lower readiness) than DR Congo. The levels of use and readiness of the complementary innovations were similar in both

Scaling readiness diagnosis of flash drying partners – DCR – May 2019

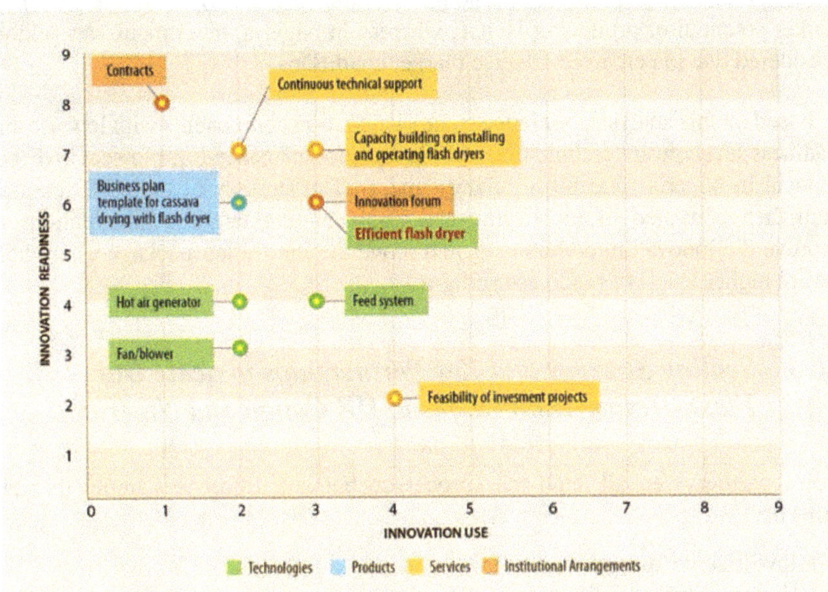

Scaling readiness diagnosis of flash drying partners – Nigeria – May 2019

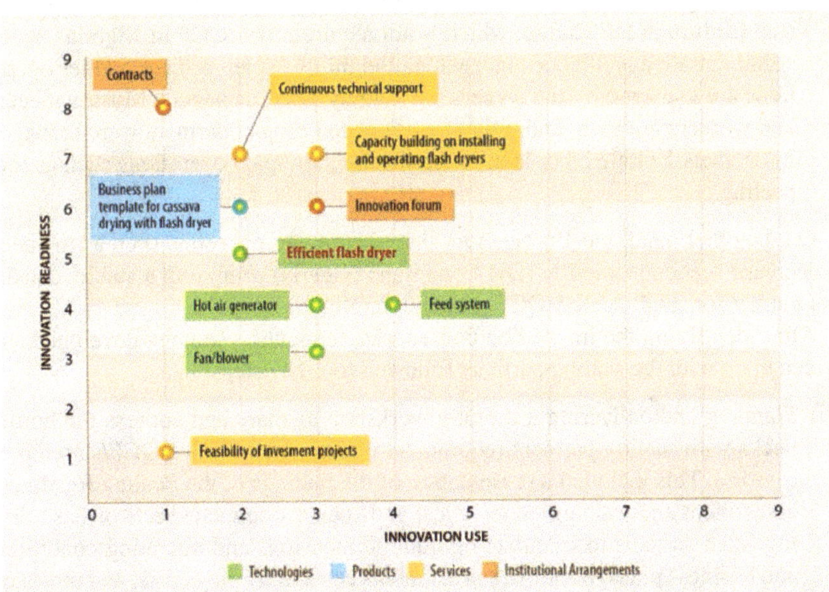

Fig. 4.8 Innovation Readiness and Innovation Use of the cassava flash-dryer innovation package in DR Congo and Nigeria after the Scaling Readiness assessment (May 2019). Core innovation components (technologies) are presented in green and complementary innovations in blue (products), yellow (services), and orange (institutional arrangements)

countries, except the feasibility of investment projects: in DR Congo, processors evaluated feasibility frequently (although with informal tools) due to the perceived market potential of producing HQCF, whereas in Nigeria, investment was seldom considered due to perceived adverse market conditions.

Based on this analysis, bottlenecks, i.e., innovation components with low use and readiness levels, were technical in DR Congo (hot air generator, blower) and economical in Nigeria (feasibility of investment). The strategy of the RTB research team then consisted of identifying and implementing targeted interventions to improve the innovation components and move the innovation package as a whole toward higher levels of readiness and use.

4.6.3 Scaling Strategy and Key Partnerships to Scale Out Flash-Drying Innovations to DR Congo and Nigeria

Project partners were all small-scale processors but with differences among them, in terms of:

(i) Investment capacity.
(ii) Factory setup, i.e., different models of machines and management of operations.
(iii) Quality of the final product expected by customers, which requires different processing techniques prior to drying. For example, lactic fermentation is a key step for fufu in DR Congo, which is not required for HQCF in Nigeria.
(iv) Administrative and labor management: eight of the ten project partner operations are managed by the owners and two by hired managers. Most managers and workers are men, and only two out of ten project partners were managed by women. Female labor is used for specific, low-paid operations such as root peeling.

On the other hand, most processors shared common concerns, such as the need to decrease fuel consumption (DR Congo and Nigeria) or finding a stable, consistent market for small-scale HQCF production (Nigeria).

After identifying the innovation bottlenecks, the scaling strategy developed and agreed upon with the scaling partners followed four key steps:

(i) Planning and delivering a training workshop to share and address the bottlenecks with scaling partners (equipment manufacturers and HQCF/starch processors). This included a presentation of the findings of the Scaling Readiness assessment and training on technical and socioeconomic aspects of flash drying, such as tools to calculate optimum dimensions and operating conditions and business plans to estimate investments costs, operating costs, and return on investment.
(ii) Implementation of the business plans by the scaling partners willing and able to do so. The process involved back-and-forth exchanges between cassava processors and the project team to develop business plans adapted to each partner's

circumstances. Realistic investment costs were provided by the equipment manufacturers who participated in the workshop. Some of the processors then used the resulting business plans to present their investment projects to banks or private investors. For those who invested their own money, the business plan contributed to accurate evaluation and of planning of the investment.

(iii) Design, construction, and commissioning of flash dryers by the scaling partners with continuous support from project scientists, through on-line technical consultations and regular on-site visits.

(iv) End-of-project debriefing workshop and evaluation of outcomes and lessons learned.

4.6.4 *Results and Outcomes of Scaling Out Flash-Drying Innovations to DR Congo and Nigeria*

By its completion in December 2020, the RTB Scaling project on flash-drying innovations fostered investment in energy-efficient, small-scale flash dryers by seven cassava processors: five in DR Congo and two in Nigeria. In addition, two demonstration flash dryers were built for promotion and training at two R&D institutions: the Federal Institute of Industrial Research Oshodi (FIIRO) in Nigeria and the International Institute of Tropical Agriculture (IITA) in Eastern Congo. Five equipment manufacturers were involved in constructing these improved flash dryers and are now well positioned with practical experience to build further dryers as new investors come forward in the future.

In terms of efficiency gains, performance evaluation of the improved flash dryers that have reached commercial operation indicated a 23 to 50% increase in production capacity (from 10 to 12–15 tons of flour/month) and at the same time a 30 to 33% reduction in fuel (diesel) consumption per kg of product, compared to the situation before the project. Feedback from the flash-dryer operators in DR Congo after incorporating the RTB innovations resulted in much improvement in drying efficiency: A processor reported that re-setting the heat exchanger burner and air inlet decreased the fuel consumption by 30% while other adjustments suggested by the project team resulted in an increase in the flour output from 150 to 250 kg/h. As a result, cassava processors reported an increase in net profitability of 8% to 10%.

Through end-of-project evaluation and learning events, the innovation package was evaluated a second time, showing the changes achieved during the project (Fig. 4.9). As intended initially, technical components moved toward higher levels of use and readiness, which reflects the focus of the research team and scaling partners to construct more energy-efficient flash dryers and prioritize the success of the investments. Complementary innovations also progressed toward higher levels of use and readiness, reflecting the positive contribution of the various events organized during the project and the continuous engagement with manufacturers and processors. Some components did not move, either because they were already at

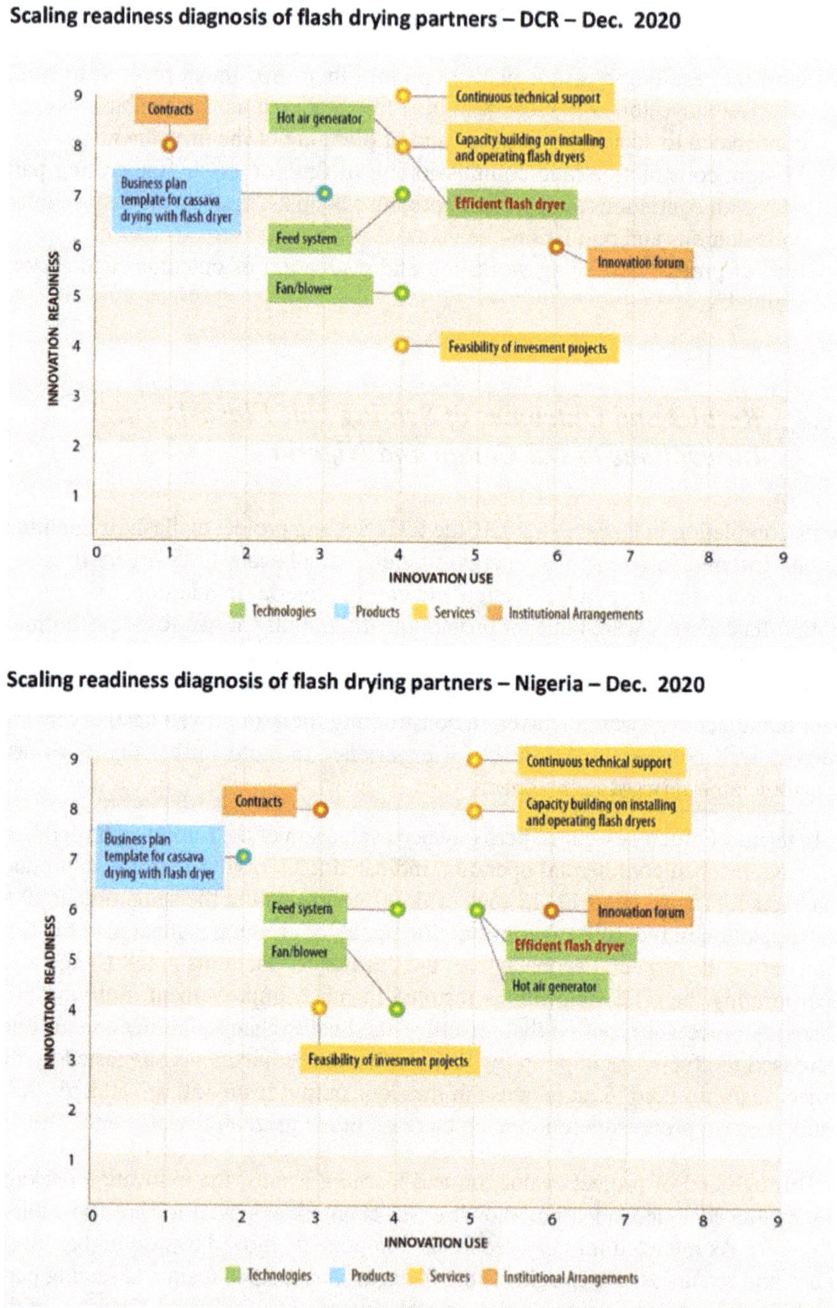

Fig. 4.9 Innovation Readiness and Innovation Use of the cassava flash-dryer innovation package in DR Congo and Nigeria at the end of the RTB Scaling flash-drying project (December 2020). Core innovation components (technologies) are presented in green and complementary innovations in blue (products), yellow (services), and orange (institutional arrangements)

sufficiently high levels of use and readiness or because of logistical delays in implementing activities, in part compounded by Covid restrictions.

Technical components reached higher levels of readiness in DR Congo compared to Nigeria. In DR Congo, the expanding market demand for HQCF to replace conventional sun-dried cassava flour boosted the determination of scaling partners to invest and to rapidly move energy-efficient flash dryers into commercial use. In contrast, in Nigeria, investment was impeded by the following:

(i) Mismatch between production capacity of HQCF factories (~1 t HQCF/day) and the purchase needs of large food companies (30 or 60 t per order, i.e., one or two trailer trucks).
(ii) Projected production costs of HQCF after adoption in flash-drying innovations may still be too high to compete easily with flour from imported wheat.
(iii) Cassava flour is not a major staple food in Nigeria, so there is limited preexisting market where HQCF could replace a traditional cassava flour by offering better quality or food safety.

During project implementation, several constraints to the development of cassava processing were identified. These included access to investment capital, stable access to cassava roots at cost-effective prices, and availability of engineering skills to conduct maintenance and repairs in a timely manner.

Over several years, RTB has made significant investments in cassava postharvest processing, first in R&D and then scaling-out activities. The outcomes of the scaling project for efficient small-scale flash drying confirm that this initiative has started to bear fruits, with flash-dryer innovations now reaching the stage of independent adoption and dissemination by the private sector in several countries.

Postharvest processing is an important part of sustainable cassava value chains to reduce losses of perishable cassava roots through transformation into food products with a long shelf life. The successful approach presented above for the efficiency of small-scale flash drying can be extended and replicated to optimize other unit operations and reduce processing costs. At the same time, more efficient processing can reduce the environmental footprint of cassava agro-industries through lower fuel and water consumption and lower product losses during processing.

4.7 Performance Diagnosis of Small-Scale Processes to Support Scaling Out of Innovations for Cassava-Based Products

Many of the activities for scaling-out innovations in cassava processing must start with an accurate comprehensive diagnosis including processing parameters, socioeconomic feasibility, and environmental impacts as a basis for decision-making for all subsequent actions. Comprehensive diagnosis is also useful at the end of scaling-out projects, as part of Monitoring, Evaluation, and Learning (MEL) to document

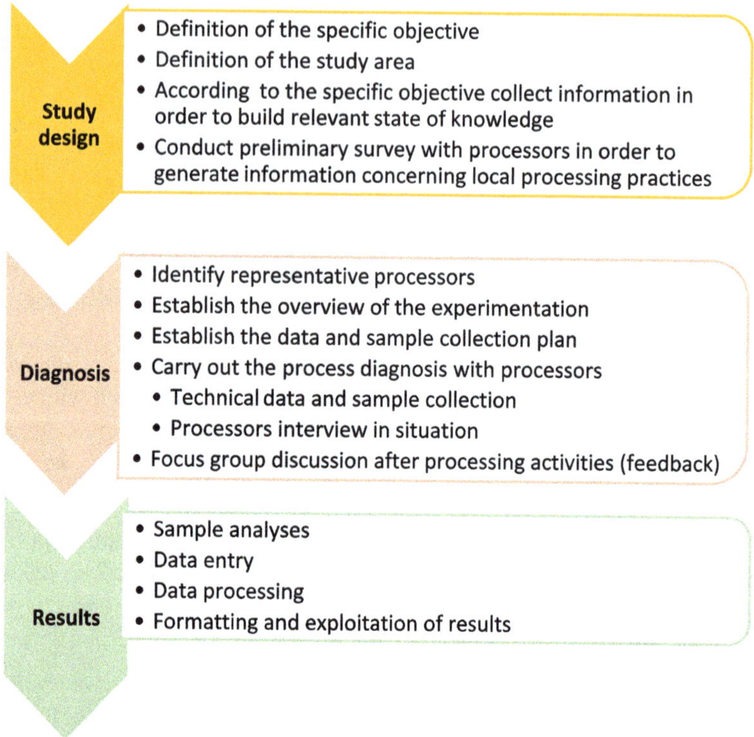

Fig. 4.10 Steps of the method for diagnosis of small-scale cassava processing units

and measure outcomes and impacts. The diagnosis contributes to improvements in product quality through the optimization of existing processes and the introduction of new operations or equipment. In this section, we lay out a versatile diagnosis method (Fig. 4.10) developed from several case studies (Adinsi et al. 2019; Escobar et al. 2018; Bouniol et al. 2017a, b, 2018, 2019, 2020) for processing of cassava into gari, fufu, and other products in West Africa, with support from RTB.

4.7.1 Design of the Diagnosis Study: Specific Objective, Study Area, and State of Knowledge

The diagnosis of small-scale processes can focus on various aspects, such as the following:

(i) Influence of cassava variety and processing on product quality.
(ii) Resource efficiency (energy, water, raw materials) in relation to production costs and environmental impacts.

(iii) Comparing different technologies for processing the same product. The first step is therefore to clearly define the specific objective(s) (Fig. 4.10), which in turn will guide the diagnosis methods to be selected.

Traditional, small-scale processing of cassava follows diverse processing pathways linked to different technologies and know-how. In turn, this variability results in different end-product quality, usually matching different consumer preferences. To collect representative data about a given process, it is therefore necessary to define where to carry out the diagnosis. These can be the areas where the target product is most frequently processed (usually areas with high cassava production), or the project's region of intervention. The diagnosis may target user segments to explicitly be socially inclusive or to take gender into account, besides other criteria: size of towns or cities, production capacity, distances to markets, etc.

The diagnosis design may be based on a review literature about the product, process, and value chain and preliminary interviews with experts, extension officers, and processors to gather information on supply, availability, and quality criteria of raw materials. The design should include a description of each of the process steps and their importance for the quality of the product, as well as gender aspects, business environment (access to infrastructures such as energy, water, road network), and market aspects.

4.7.2 Diagnosis

Conducting a process diagnosis requires time, qualified people, and funding, so preplanning is important (Sect. 3.1) to coordinate the collection of samples and data and to ensure the quality of the results.

Identify representative processors The first step of the diagnosis is to identify in the study area a panel of processors representative of current practices, according to the following criteria:

- Main economic activity based on the studied product.
- Recognized know-how and ability to produce a specific quality of finished products matching the expectations of consumers.
- Gender division of labor among the selected processors that is representative of actual gender labor distribution.
- Technological level among the selected processors that is representative of the local most common technology. If most processors are small scale, the focus should be on them to mitigate negative social consequences of scaling technologies and to ensure more users of the technology from a social inclusive perspective.
- Evaluate the volumes processed (by month/week/year) and the type of market targeted (retail, wholesale, town, village, etc.).

Overview of the experimentation A *schematic overview of the experimentation* is necessary to clearly delimit the scope of the diagnosis according to the specific objective, to let the experimenters understand where they intervene in the overall process, and to refine the diagnosis by ensuring that the workflow and experimental design are suitable to collect scientifically valid data (representativeness of the sampling, number of repetitions). Several approaches are possible:

(i) Replicate the entire transformation process from the same batch of raw cassava several times. Tens or hundreds of kilograms are often necessary.
(ii) Repeat the experimentation over several years. As much as possible, use the same cassava varieties harvested at the same age and processed under the same conditions.
(iii) If appropriate, repeat only specific unit operations, for example, in the case of a series of complex unit operations where all variables may be difficult to control at the same time.

Data and sample collection plan The *collection plan* (e.g., Table 4.6) maps all the data and samples to collect at each step of the process, which may include temperature, weight, flow rate, dry matter, relative humidity, time, pH, and pressure. This helps to manage the complexity of the diagnosis by checking in advance with the processors the feasibility of the planned measurements and by keeping track of data and sample collection during the diagnosis.

Carry out the process diagnosis with processors

Technical data collection The same batch of *raw materials* (e.g., cassava roots) should be used for all the replications and in all the locations of the diagnosis to

Table 4.6 Example of data collection plan: diagnosis of "bâtons de manioc" (cassava sticks) in study from Cameroon in 2015

Data and sample acquisition	Weight; material balance	Sample collect. / Lab. analyses	Dry matter	Time	Work force	Water, pH, energy	Temperature	Losses of dry material	Remarks
Raw material	X Roots		X						Photos of outside and inside of roots
Peeling	X Roots & peels			X	X				Photos
Washing				X	X	X Volume water			
Steeping / Rinsing	X Initial weight X Final weight	X Water end X Filtered water	X Whatman X pulp	X		X Volume water X Final water		X Whatman filter X Filtered water	Daily observations with operator advisories. Photos
Draining / Grinding	X Final weight	X	X	X	X				Photos
Crushing	X	X Crushed pulp		X	X	X Type of motor, power			Principle & dimensions of crushing system. Rotation speed. Photos
Shaping	X Leaves X Bâtons	X Leaves							Sampling sheets for drying. Photos
Cooking	X End cooking	X kinetics	X	X	X	X Weight wood or gas	X DAQ (Almemo) X + cooling		Principle & dimensions of cooking system. Photos
Bâtons	X	X							Photos

facilitate comparisons between technologies. This may not always be possible, however, due to the short shelf life of fresh roots and transportation distances. In this case, raw materials are sourced locally in each location, and the raw material is considered as one variable of the experimentation. Raw materials then need to be described in detail, including photos of the fresh roots (before peeling), size measurements, and any appropriate laboratory analyses.

All *equipment, instruments, and data collection sheet* necessary to carry out the experiments, to collect samples, and to record data are prepared in advance, based on the collection plan. The following characterization protocol is then applied: Before the first unit operation, the research team records the weight of the raw materials that will be processed into the product under study. Then after each unit operation (peeling, washing, etc.), the duration is recorded and the intermediate product (e.g., peeled product) and residues (e.g., peels, stems, soil) are weighed for later calculations of yields and mass balance. Samples can also be collected for further laboratory analyses (dry matter, starch content, fibers, pH, etc.). The following diagnosis parameters are then calculated for each unit operation and the process as a whole:

- *Yield*, defined as the quantity of product recovered after each unit operation and expressed as percentage (wet basis) of the quantity of raw material.
- *Material balance* evaluated by checking that the weights of all the inputs (raw materials, water, etc.) and outputs (final product, peels, fibers, wastewater, etc.) are equal.
- *Productivity*, defined as the quantity of raw material (in kg) processed per hour and per processor. This is an important indicator of efficiency and also of drudgery.

Interview of processors and participative approach To capture processors' expertise, at each step of the process, interviews can elicit the details that processors use to recognize that the raw materials or intermediate products will give a good final product and how they may adjust their process to ensure the best possible final product. In the case of experiments comparing the processing potential of several varieties, processors can be asked to rank them from good to poor. This ranking can be repeated with intermediate products during processing.

4.7.3 Exploitation of Results

In addition to regular reporting of results collected for each unit operation, the process as a whole can also be analyzed with the following:

- The *process flowsheet* (e.g., Fig. 4.11) describes the sequence of the unit operations, their duration (productivity), inputs and outputs (raw materials, water, electricity, thermal energy, etc.), and material balance based on the yields of the intermediate products.

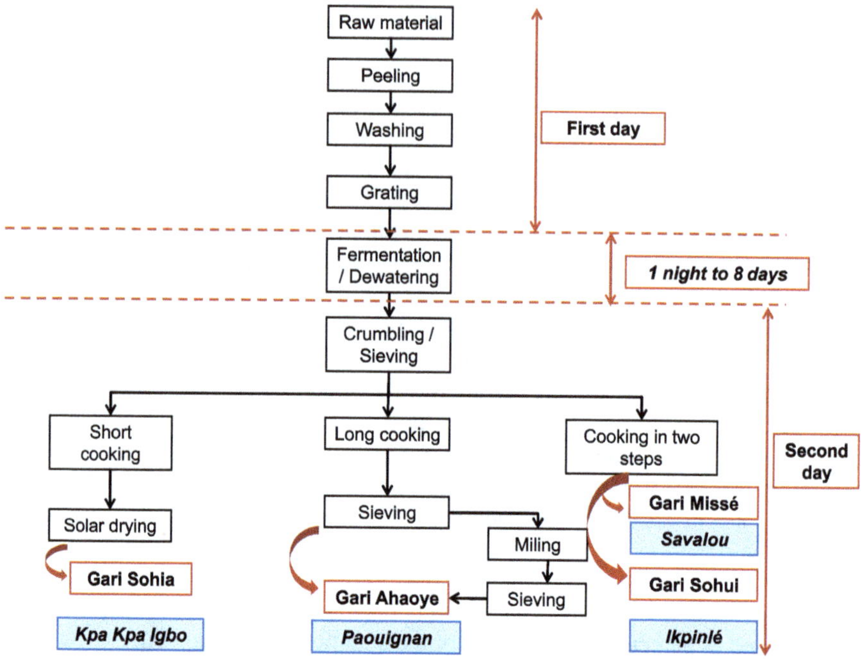

Fig. 4.11 Example of a process diagram for gari processing in four locations (shown in blue, italics) (study from Benin in 2015)

– *Yields and productivity of intermediate products*: The yields of intermediate products reveal which unit operations most impact the overall yield of the process (e.g., Fig. 4.12). Productivity of each unit operation reveals bottlenecks, i.e., which operations slow down the overall process. Yields and productivity can also depend on the processing ability of different cassava varieties, as some varieties can be harder to process, e.g., more difficult to peel, thicker peels, lower dry matter, or containing fibers that need to be manually removed. Yield and productivity are important acceptability criteria that influence the adoption of improved technologies and/or new varieties, in addition to acceptability criteria of the end product such as visual appearance, texture, and taste.

4.7.4 Conclusions

Process diagnosis is most effective when implemented with a formal methodological framework such as the one presented above. Combined with socioeconomic and environmental surveys, process diagnosis is a prerequisite to scaling-out innovations for several purposes: to provide a baseline to monitor the progress and outcomes of scaling-out activities; to reveal local constraints and expectations for the

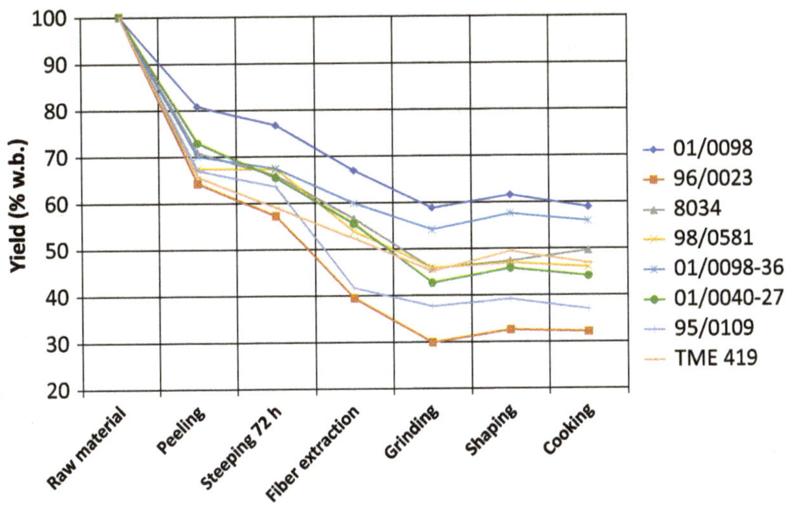

Fig. 4.12 Example of the yields of the unit operations for processing gari. Eight cassava varieties revealed the effect of variety on process yields (study from Cameroon in 2015)

innovations with respect to local traditions and gender, employment, and environmental and economic criteria; and to identify the innovation components package and establish the scaling readiness diagram when applying the Scaling Readiness framework. A well-executed diagnosis provides not only reliable information on the process and product under study but also builds dialog with processors on the benefits and constraints of processing technologies and improved cassava varieties to facilitate the adoption of innovations.

4.8 Conclusions and Perspectives: Ongoing Research and Strategic Areas for Future Research on Cassava Processing

Since 2013, RTB has fostered several initiatives to improve cassava processing and federated a team of multidisciplinary researchers and engineers from IITA, CIAT, CIRAD, NRI, academic partners, and private partners in several countries. One notable outcome has been the reengineering of small-scale flash dryers for better energy efficiency and lower production costs. The flash-drying results presented in this chapter are useful to illustrate in a practical, step-by-step way the overall approach to reengineering postharvest technologies, from the initial state-of-the-art surveys to the stages of design, proof of concept, pilot testing, and scaling out with private partners. This work also provided a relevant case study for the application of Scaling Readiness, showing the progression of the project through the levels of innovation readiness and use (Table 4.5).

This experience shows that R&D interventions at postharvest processing level have an important part to play in advancing sustainable value chains of RTB crops, especially when combined with scaling-out activities in partnership with private small-scale processors and equipment manufacturers. In addition to flash dryers, several cassava products and unit operations have been investigated by RTB: cabinet drying (Precoppe et al. 2017), attiéké (Alamu et al. 2020), gari (Escobar et al. 2018, 2021; Dahdouh et al. 2021), as well as links between processing and human health (Parmar et al. 2019; Bede-Ojimadu and Orisakwe 2020). Applying the same reengineering approach to other unit operations such as peeling, grating, dewatering, and milling will improve the overall efficiency of small-scale cassava industries and over time should increase incomes and employment along the whole cassava value chain (Escobar et al. 2021; Ezeocha et al. 2019; Dahdouh et al. 2021; Dou et al. 2020). Apart from interventions on postharvest processing, RTB researchers are currently also working on mobile cassava processing units to reduce the time and distance from the field to final products, so as to address the barrier of poor transportation infrastructure in West Africa.

The perspectives for cassava value chains over the next 10–20 years are an overall expansion to meet the food security and nutritional needs of growing populations. In this context, gains in processing efficiency will be crucial to improving both product quality and production capacity, which in turn are a prerequisite to connect rural small-scale cassava processors with higher-value and more distant markets, in particular large cities and international exports. Integrating processing innovations with the current scale of processing is important to provide a comparative advantage from the bottom up, as focusing on larger-scale processing can result in scale advantages and also put smaller-scale processors out of business with the risk of increasing inequality, social instability, and adverse gender effects as women are often highly represented in small-scale processing.

Acknowledgments This research was undertaken as part of, and funded by, the CGIAR Research Program on Roots, Tubers and Bananas (RTB) and supported by CGIAR Trust Fund contributors. Funding support for this work was provided by CAVA-II, CIRAD, and Technologies for African Agricultural Transformation (TAAT) Cassava Compact. The authors thank all the data providers, in particular cassava processors and equipment manufacturers, who have kindly participated to this study, as well as the companies Nutripro, Agrimac, Layuka, and Ecosac (DR Congo); Open Door System, Hickman Ventures, Deban Faith, Lentus Food, and El-Rasheed Farms (Nigeria); Angavil (Dominican Republic); Codipsa (Paraguay); and Deriyuca (Colombia). The participation of Charlène Lancement, Aymeric Delafosse, Francisco Giraldo, Andrés Escobar, Jhon Larry Moreno, Jorge Luna, and María Alejandra Ospina in support of project activities is gratefully acknowledged.

References

Abass AB, Mlingi N, Ranaivoson R, Zulu M, Mukuka I, Abele S, Bachwenkizi B, Cromme N (2013) Potential for commercial production and marketing of cassava: experiences from the small-scale cassava processing project in East and Southern Africa. IITA, Ibadan

Abass AB, Amaza P, Bachwenkizi B, Alenkhe B, Mukuka I, Cromme N (2017) Adding value through the mechanization of postharvest cassava processing, and its impact on household

poverty in north-eastern Zambia. Appl Econ Lett 24(9):579–583
Abass AB, Awoyale W, Alenkhe B, Ndavi M, Asiru BW, Manyong V, Sanginga N (2018) Can food technology innovation change the status of a food security crop? A review of cassava transformation into "bread" in Africa. Food Rev Intl 34(1):87–102
Adegbite SA, Abass AB, Olukunle OJ, Olalusi AP, Asiru WB, Awoyale W (2019) Mass and energy balance analysis of pneumatic dryers for cassava and development of optimization models to increase competitiveness in Nigeria. Curr J Appl Sci Technol 32(3):1–11. https://doi.org/10.9734/CJAST/2019/46215
Adenle AA, Manning L, Azadi H (2017) Agribusiness innovation: a pathway to sustainable economic growth in Africa. Trends Food Sci Technol 59:88–104. https://doi.org/10.1016/j.tifs.2016.11.008
Adinsi L, Akissoé N, Escobar A, Prin L, Kougblenou N, Dufour D, Hounhouigan DJ, Fliedel G (2019) Sensory and physicochemical profiling of traditional and enriched gari in Benin. Food Sci Nutr 7(10):3338–3348
Aichayawanich S, Nopharatana M, Nopharatana A, Songkasiri W (2011) Agglomeration mechanisms of cassava starch during pneumatic conveying drying. Carbohydr Polym 84:292–298
Alamu EO, Ntawuruhunga P, Chibwe T, Mukuka I, Chiona M (2019) Evaluation of cassava processing and utilization at household level in Zambia. Food Secur 11:141–150
Alamu EO, Abass A, Maziya-Dixon B, Diallo TA, Sangodoyin MA, Kolawole P, Tran T, Awoyale W, Kulakow P, Parkes E, Kouame KA, Amani K, Appi A, Dixon A (2020) Report on the status of Attiéké production in Côte d'Ivoire. International Institute of Tropical Agriculture (IITA), Ibadan, p 42
Alonso L, Viera MA, Best R, Gallego S, García JA (2012) Artificial cassava drying systems. In: Ospina Patiño B, Ceballos H (eds) Cassava in the third millennium: modern production, processing, use, and marketing systems. Centro Internacional de Agricultura Tropical (CIAT); Latin American and Caribbean Consortium to support Cassava Research and Development (CLAYUCA); Technical Center for Agricultural and Rural Cooperation (CTA), Cali, pp 427–441
AOAC (1998) Moisture in malt (method 935.29). In: Official methods of analysis. Association of Official Analytical Chemists (AOAC), Gaithersburg
Awoyale W, Abass AB, Ndavi M, Maziya-Dixon B, Sulyok M (2017) Assessment of the potential industrial applications of commercial dried cassava products in Nigeria. J Food Meas Charact 11:598–609
Bede-Ojimadu O, Orisakwe OE (2020) Exposure to wood smoke and associated health effects in Sub-Saharan Africa: a systematic review. Ann Glob Health 86(1):32
Bouniol A, Prin L, Hanna R, Fotso A, Fliedel G (2017a) Assessment of the processability of improved cassava varieties into a traditional food product ("baton" or "chikwangue") in Cameroon. CGIAR Research Program on Roots, Tubers and Bananas (RTB), Lima, Peru. RTB working paper, p 28
Bouniol A, Ospina MA, Fotso A, Hanna R, Dufour D (2017b) Assessment of the processing ability of improved cassava varieties into a traditional food product (fufu) in Cameroon. CGIAR Research Program on Roots, Tubers and Bananas (RTB), Lima, Peru. RTB working paper, p 24
Bouniol A, Escobar A, Adinsi L, Prin L, Fliedel G, Tran T, Dufour D (2018) Standard Operating Procedure (SOP) to establish a diagnosis of a process with processors. CGIAR Research Program on Roots, Tubers and Bananas (RTB), Lima, Peru. RTB working paper, p 25
Bouniol A, Adinsi L, Padonou SW, Hotegni F, Gnanvossou D, Akissoé N, Fliedel G, Tran T, Dufour D (2019) Assessment of the processing ability of improved cassava varieties into a traditional food product (lafun) in Benin. CGIAR Research Program on Roots, Tubers and Bananas (RTB), Lima, Peru. RTB working paper, p 29
Bouniol A, Adinsi L, Hotegni F, Delpech A, Akissoé N, Fliedel G, Tran T, Dufour D (2020) Assessment of the processing ability of improved cassava varieties into a traditional food product (gari) in Benin. CGIAR Research Program on Roots, Tubers and Bananas (RTB), Lima, Peru. RTB working paper, p 34
Bouniol A, Adinsi L, Padonou SW, Hotegni F, Gnanvossou D, Tran T, Dufour D, Hounhouigan DJ,

Akissoé N (2021) Rheological and textural properties of lafun, a stiff dough, from improved cassava varieties. Int J Food Sci Technol. https://doi.org/10.1111/ijfs.14902

Brennan JG (2011) Evaporation and dehydration. In: Brennan JG, Grandison AS (eds) Food processing handbook, 2nd edn. Wiley-VCH, Weinheim, pp 77–130

Breuninger WF, Piyachomkwan K, Sriroth K (2009) Tapioca/cassava starch: production and use. In: BeMiller J, Whistler R (eds) Starch: chemistry and technology. Academic Press, San Diego, pp 541–568

Chapuis A, Precoppe M, Méot JM, Sriroth K, Tran T (2017) Pneumatic drying of cassava starch: numerical analysis and guidelines for the design of efficient small-scale dryers. Dry Technol 35:393–408

Crapiste GH, Rotstein E (1997) Design and performance evaluation of dryers. In: Valentas KJ, Singh RP, Rotstein E (eds) Handbook of food engineering practice. CRC Press, Boca Raton, pp 121–162

Da G, Dufour D, Giraldo A, Moreno M, Tran T, Vélez G, Sánchez T, Le Thanh M, Marouzé C, Maréchal PA (2013) Cottage level cassava starch processing systems in Colombia and Vietnam. Food Bioprocess Technol 6(8):2213–2222. https://doi.org/10.1007/s11947-012-0810-0

Dahdouh L, Escobar A, Rondet E, Ricci J, Fliedel G, Adinsi L, Dufour D, Cuq B, Delalonde M (2021) Role of dewatering and roasting parameters on the quality of handmade gari. Int J Food Sci Technol 56(3):1298–1310. https://doi.org/10.1111/ijfs.14745

Dou GY, Wang XT, Zhao BC, Yuan XA, Pan CX, Tran T, Zellweger H, Zhu KS, Guo YJ, Wu H, Yin J, Bai YY (2020) The transformation and outcome of traditional cassava starch processing in Guangxi, China. Environ Technol. https://doi.org/10.1080/09593330.2020.1725647

Dufour D, O'Brien GM, Best R (2002) Cassava flour and starch: progress in research and development. Centre de Coopération Internationale en Recherche Agronomique pour le Développement (CIRAD), Montpellier, France and Centro Internacional de Agricultura Tropical (CIAT), Cali, 409 pp

Dziedzoave NT, Abass AB, Amoa-Awua WKA, Sablah M (2006) In: Adegoke GO, Brimer L (eds) Quality management manual for the production of high quality cassava flour. International Institute of Tropical Agriculture (IITA), Ibadan, 68 p

Edeh JC, Nwankwojike BN, Abam FI (2020) Design modification and comparative analysis of cassava attrition peeling machine. AMA Agric Mech Asia Africa Latin America 51(1):63–71

Escobar A, Dahdouh L, Rondet E, Ricci J, Dufour D, Tran T, Cuq B, Delalonde M (2018) Development of a novel integrated approach to monitor processing of cassava roots into Gari: macroscopic and microscopic scales. Food Bioprocess Technol 11:1370–1380

Escobar A, Rondet E, Dahdouh L, Ricci J, Akissoé N, Dufour D, Tran T, Cuq B, Delalonde M (2021) Identification of critical versus robust processing unit operations determining the physical and biochemical properties of cassava-based semolina (gari). Int J Food Sci Technol 56(3):1311–1321. https://doi.org/10.1111/ijfs.14857

Ezeocha CV, Ihesie LC, Kanu AN (2019) Comparative evaluation of toasting variables and the quality of gari produced by different women in Ikwuano LGA, Abia State, Nigeria. J Food Process Preserv 43(9):e14060. https://doi.org/10.1111/jfpp.14060

Forsythe L, Martin AM, Posthumus H (2015) Cassava market development: a path to women's empowerment or business as usual? Food Chain 5(1–2):11–27. https://doi.org/10.3362/2046-1887.2015.003

Forsythe L, Posthumus H, Martin AM (2016) A crop of one's own? Women's experiences of cassava commercialization in Nigeria and Malawi. J Gender Agric Food Secur 1(2):110–128

Hansupalak N, Piromkraipak P, Tamthirat P, Manitsorasak A, Sriroth K, Tran T (2016) Biogas reduces the carbon footprint of cassava starch: a comparative assessment with fuel oil. J Clean Prod 134(part B):539–546

Kemp IC (2012) Fundamentals of energy analysis of dryers. In: Tsotsas E, Mujumdar AS (eds) Modern drying technology: energy savings, vol 4. Wiley-VCH, Weinheim, pp 1–45

Kitinoja L, Saran S, Roy SK, Kader AA (2011) Postharvest technology for developing countries: challenges and opportunities in research, outreach and advocacy. J Sci Food Agric 91(4):597–603. https://doi.org/10.1002/jsfa.4295

Kudra T (2009) Energy aspect in food dehydration. In: Ratti C (ed) Advances in food dehydration. CRC Press, Boca Raton, pp 423–445

Kussaga JB, Jacxsens L, Tiisekwa BPM, Luning PA (2014) Food safety management systems performance in African food processing companies: a review of deficiencies and possible improvement strategies. J Sci Food Agric 94(11):2154–2169. https://doi.org/10.1002/jsfa.6575

Kuye A, Ayo DB, Sanni LO, Raji AO, Kwaya EI, Otuu OO, Asiru WB, Alenkhe B, Abdulkareem IB, Bamkefa B, Tarawali G, Dixon AGO, Okechukwu RU (2011) Design and fabrication of a flash dryer for the production of high quality cassava flour. Cassava Enterprise Development Project, International Institute of Tropical Agriculture, Ibadan, 50 pp

Luna J, Dufour D, Tran T, Pizarro M, Calle F, Garcia Dominguez M, Hurtado IM, Sanchez T, Ceballos H (2021) Postharvest physiological deterioration in several cassava genotypes over sequential harvests and effect of pruning prior to harvest. Int J Food Sci Technol 56(3):1322–1332. https://doi.org/10.1111/ijfs.14711

Mujumdar AS (2006) Handbook of industrial drying, 3rd edn. CRC Press, Taylor & Francis Group, Boca Raton, 1312pp

Ndjouenkeu R, Kegah FN, Teeken B, Okoye B, Madu T, Olaosebikan OD, Chijioke U, Bello A, Osunbade AO, Owoade D, Takam-Tchuente NH, Njeufa EB, Nguiadem-Chomdom ILN, Forsythe L, Maziya-Dixon B, Fliedel G (2021) From cassava to gari: mapping of quality characteristics and end-user preferences in Cameroon and Nigeria. Int J Food Sci Technol 56(3):1223–1238. https://doi.org/10.1111/ijfs.14790

Nweke FI (1994) Cassava processing in Sub-Saharan Africa: the implications for expanding cassava production. Outlook Agric 23(3):197–205. https://doi.org/10.1177/003072709402300307

Nzudie HLF, Zhao X, Tillotson MR, Zhang F, Li YP (2020) Modelling and forecasting roots & tubers losses and resulting water losses in sub-Saharan Africa considering climate variables. Phys Chem Earth 120:102952. https://doi.org/10.1016/j.pce.2020.102952

Ojide M, Abass A, Adegbite S, Tran T, Taborda LA, Chapuis A, Lukombo S, Totin E, Sartas M, Schut M, Becerra Lopez-Lavalle LA, Dufour D (2021) Processors' experience in the use of flash dryer for cassava-derived products in Nigeria. Frontiers in Sustainable Food Systems. https://www.frontiersin.org/articles/10.3389/fsufs.2021.771639/abstract

Okudoh V, Trois C, Workneh T, Schmidt S (2014) The potential of cassava biomass and applicable technologies for sustainable biogas production in South Africa: a review. Renew Sustain Energy Rev 39:1035–1052. https://doi.org/10.1016/j.rser.2014.07.142

Oni KC, Oyelade OA (2014) Mechanization of cassava for value addition and wealth creation by the rural poor of Nigeria. AMA Agric Mech Asia Africa Latin America 45(1):66–78

Ozoegwu CG, Eze C, Onwosi CO, Mgbemene CA, Ozor PA (2017) Biomass and bioenergy potential of cassava waste in Nigeria: estimations based partly on rural-level garri processing case studies. Renew Sustain Energy Rev 72:625–638. https://doi.org/10.1016/j.rser.2017.01.031

Pakowski Z, Mujumdar AS (2014) Drying of pharmaceutical products. In: Mujumdar AS (ed) Handbook of industrial drying. CRC Press, Boca Raton, pp 681–701

Parmar A, Tomlins K, Sanni L, Omohimi C, Thomas F, Tran T (2019) Exposure to air pollutants and heat stress among resource-poor women entrepreneurs in small-scale cassava processing. Environ Monit Assess 191:693

Patrizi N, Bruno M, Saladini F, Parisi ML, Pulselli RM, Bjerre AB, Bastianoni S (2020) Sustainability assessment of biorefinery systems based on two food residues in Africa. Front Sustain Food Syst 4:522614. https://doi.org/10.3389/fsufs.2020.522614

Precoppe M, Tran T, Chapuis A, Müller J, Abass A (2016) Improved energy performance of small-scale pneumatic dryers used for processing cassava in Africa. Biosyst Eng 151:510–519

Precoppe M, Chapuis A, Müller J, Abass A (2017) Tunnel dryer and pneumatic dryer performance evaluation to improve small scale cassava processing in Tanzania. J Food Process Eng 40:1–10

Precoppe M, Komlaga GA, Chapuis A, Müller J (2020) Comparative study between current practices on cassava drying by small-size enterprises in Africa. Appl Sci 10:7863

Romuli S, Abass A, Müller J (2017) Physical properties of Cassava Grits before and after pneumatic drying. J Food Process Eng 40:e12397

Saengchan K, Tran T, Faye M, Cantero-Tubilla B (2015) Technical and economic assessment

of cassava starch processing in Paraguay. CGIAR Research Program on Roots, Tubers and Bananas (RTB), Lima, Peru. RTB working paper, p 52
Saravacos G, Kostaropoulos AE (2016) Food evaporation equipment. In: Handbook of food processing equipment. Springer, Cham, pp 367–419
Sartas M, Schut M, Proietti C, Thiele G, Leeuwis C (2020a) Scaling readiness: science and practice of an approach to enhance the impact of research for development. Agric Syst 183:102874
Sartas M, Schut M, van Schagen B, Velasco C, Thiele G, Proietti C, Leeuwis C (2020b) Scaling readiness concepts, practices and implementation. International Potato Center on behalf of RTB, 217 p
Sriroth K, Piyachomkwan K, Wanlapatit S, Oates CG (2000) Cassava starch technology: the Thai experience. Starch-Stärke 52(12):439–449
Standard B (2004) Guide to the measurement of humidity (Standard BS 1339-3:2004). British Standards Institution, London
Strumiłło C, Jones PL, Żyłła R (2014) Energy aspects in drying. In: Mujumdar AS (ed) Handbook of industrial drying. CRC Press, Boca Raton, pp 1077–1100
Taborda LA (2018) Determinación y análisis integral de impactos de la agroindustria rural de almidón de yuca en Cauca, Colombia. Doctoral thesis, Universidad Nacional de Colombia - Palmira
Taiwo KA (2006) Utilization potentials of cassava in Nigeria: the domestic and industrial products. Food Rev Intl 22(1):29–42. https://doi.org/10.1080/87559120500379787
Taiwo KA, Fasoyiro S (2015) Women and cassava processing in Nigeria. Int J Dev Res 5(2):3513–3517
Teeken B, Olaosebikan O, Haleegoah J, Oladejo E, Madu T, Bello A, Parkes E, Egesi C, Kulakow P, Kirscht H, Tufan H (2018) Cassava trait preferences of men and women farmers in Nigeria: implications for breeding. Econ Bot 72:263–277
Teeken B, Agbona A, Bello A, Olaosebikan O, Alamu E, Adesokan M, Awoyale W, Madu T, Okoye B, Chijioke U, Owoade D, Okoro M, Bouniol A, Dufour D, Hershey C, Rabbi I, Maziya-Dixon B, Egesi C, Tufan H, Kulakow P (2021) Understanding cassava varietal preferences through pairwise ranking of gari-eba and fufu prepared by local farmer-processors. Int J Food Sci Technol 56(3):1258–1277. https://doi.org/10.1111/ijfs.14862
Thiele G, Dufour D, Vernier P, Mwanga ROM, Parker ML, Schulte Geldermann E, Teeken B, Wossen T, Gotor E, Kikulwe E, Tufan H, Sinelle S, Kouakou AM, Friedmann M, Polar V, Hershey C (2021) A review of varietal change in roots, tubers and bananas: consumer preferences and other drivers of adoption and implications for breeding. Int J Food Sci Technol 56(3):1076–1092. https://doi.org/10.1111/ijfs.14684
Tran T, Da G, Moreno-Santander MA, Velez-Hernandez GA, Giraldo-Toro A, Piyachomkwan K, Sriroth K, Dufour D (2015) A comparison of energy use, water use and carbon footprint of cassava starch production in Thailand, Vietnam and Colombia. Resour Conserv Recy 100:31–40. https://doi.org/10.1016/j.resconrec.2015.04.007
Tran T, Faye M, Hansupalak N, Cantero-Tubilla B (2015a) Technical and economic assessment of cassava starch processing in Thailand. CGIAR Research Program on Roots, Tubers and Bananas (RTB), Lima, Peru. RTB working paper, p 65
Tran T, Faye M, Cantero-Tubilla B (2015b) Technical and economic assessment of cassava starch processing in Northern Vietnam. CGIAR Research Program on Roots, Tubers and Bananas (RTB), Lima, Peru. RTB working paper, p 59
Tran T, Faye M, Cantero-Tubilla B (2015c) Technical and economic assessment of cassava starch processing in Colombia. CGIAR Research Program on Roots, Tubers and Bananas (RTB), Lima, Peru. RTB working paper. 59 pp
UNEP: United Nations Environment Programme (2007) Global environment outlook GEO4: environment for development. UNEP, Nairobi
Yank A, Ngadi M, Kok R (2016) Physical properties of rice husk and bran briquettes under low pressure densification for rural applications. Biomass Bioenergy 84:22–30. https://doi.org/10.1016/j.biombioe.2015.09.015

Zvinavashe E, Elbersen HW, Slingerland M, Kolijn S, Sanders JPM (2011) Cassava for food and energy: exploring potential benefits of processing of cassava into cassava flour and bioenergy at farmstead and community levels in rural Mozambique. Biofuels Bioprod Biorefining Biofpr 5(2):151–164. https://doi.org/10.1002/bbb.272

5

OFSP Puree: Development, Scaling and Commercialization

Mukani Moyo, Van-Den Truong, Josip Simunovic, Jean Pankuku, George Ooko Abong, Francis Kweku Amagloh, Richard Fuchs, Antonio Magnaghi, Srinivasulu Rajendran, Fredrick Grant, and Tawanda Muzhingi

Abstract Replacing some of the wheat flour in breads and pastries with OFSP (orange-fleshed sweetpotato) puree can increase the market demand for these nutritious varieties and would offer economic opportunities for smallholders, including women and youths. The technology to make sweetpotato puree has been well developed in industrialized countries since the 1960s. Techniques fine-tuned by RTB allow OFSP puree to be stored in plastic bags for 6 months, without refrigeration. Private companies in Malawi and Kenya are now manufacturing the puree and selling it to bakeries that substitute OFSP puree for up to 40% of the white wheat flour in bread and other baked goods. Consumers like the bread that is sold in supermarkets and bakeries. Food safety protocols ensure that the puree is part of safe,

M. Moyo (✉)
International Potato Center (CIP), Regional Office for Africa, Nairobi, Kenya
e-mail: Mukani.Moyo@cgiar.org

V.-D. Truong · J. Simunovic · T. Muzhingi
Department of Food, Bioprocessing and Nutritional Sciences, North Carolina State University, Raleigh, NC, USA
e-mail: vtruong@ncsu.edu; simun@ncsu.edu

J. Pankuku
Tehilah Value Chain and Food Processing, Blantyre, Malawi

Root and Tuber Crops Development Trust (RTCDT), Lilongwe, Malawi

G. O. Abong
Department of Food Science, Nutrition and Technology, University of Nairobi, Nairobi, Kenya

F. K. Amagloh
Department of Food Science & Technology, Faculty of Agriculture, Food and Consumer Sciences, University for Development Studies, Tamale, Ghana
e-mail: fkamagloh@uds.edu.gh

healthy products. The OFSP seed is available to smallholder farmers, who are linked with processors who buy the roots. Business models suggest that processing puree is profitable. The Scaling Readiness approach is helping to ensure that more farmers, processors, and consumers benefit from OFSP.

5.1 Introduction

Sweetpotato is a climate-smart root crop that adapts well to low rainfall, different altitudes, and poor soils. It requires minimal labor or other farm inputs. It is a hardy crop with low risk of crop failure (Abidin et al. 2017; Low et al. 2020). These farmer-preferred agronomic traits, combined with relatively high nutritional value, make the sweetpotato an ideal crop for enhancing food security and improving livelihoods in African communities. The biofortified orange-fleshed sweetpotato (OFSP) varieties have high levels of β-carotene, a precursor of vitamin A in the human body. OFSP is a proven, effective, and sustainable source of vitamin A, significantly contributing to the fight against vitamin A deficiency (VAD) in Africa (Low et al. 2007; Girard et al. 2017). OFSP also has vital, life-promoting phytochemicals that enhance protection from peroxides (Abong et al. 2020).

Despite the importance and knowledge of the nutritional benefits of OFSP, it remains generally underutilized, possibly because the roots are highly perishable, which depresses their market value. Low prices can subject OFSP to market losses, discouraging farmers from planting the varieties. Consumers prefer white-fleshed sweetpotato varieties, further reducing demand for OFSP (Moyo et al. 2021). The sweetpotato's seasonality further limits the periods when consumers have access to OFSP. Urban consumers eat less OFSP because they prefer convenient, ready-to-eat foods over roots that need to be cooked. Fresh sweetpotato is eaten boiled, steamed, roasted, or fried in most African communities (Mwanga et al. 2021), so processing OFSP roots into a puree creates more options for consuming them, improving their availability in cities and reducing food waste and losses.

OFSP puree can replace some of the white, wheat flour in baked and fried products. Making puree overcomes the challenge of perishability and seasonality of the roots by processing them into a more shelf-stable product that can be incorporated

R. Fuchs
Natural Resources Institute (NRI), University of Greenwich, Kent, UK
e-mail: R.S.Fuchs@greenwich.ac.uk

A. Magnaghi
Euro Ingredients Limited, Southern By-pass, Kikuyu, Kiambu County, Kenya
e-mail: antonio@euroingredients.net

S. Rajendran · F. Grant
International Potato Center (CIP), Kampala, Uganda
e-mail: srini.rajendran@cgiar.org; F.grant@cgiar.org

into diets even when the fresh roots are off-season (Wanjuu et al. 2018). Bread is a staple food in much of Africa, particularly in cities. Incorporating OFSP puree into bread would significantly increase the number of OFSP consumers and reduce VAD (Awuni et al. 2018; Owade et al. 2018). OFSP puree can replace up to 50% of the wheat flour in bread, while reducing sugar (90%) and fat (50%) and eliminating artificial colorings (egg yellow). The baked bread retains over 50% of the β-carotene, and the OFSP puree improves the texture of wheat products, making them easy to chew and digest (Wanjuu et al. 2018).

OFSP puree in bakery products is new and unique to sub-Saharan Africa (SSA) and shows great potential for expansion. African millers spend millions importing wheat, with East African countries being among the top importers. Kenya's wheat import bills are estimated at $250 million, Tanzania's at $150 million, Uganda's at $53 million, and Rwanda's at $35 million per year. Substituting some of this wheat flour with locally produced OFSP puree in popular baked goods would save African economies much needed foreign currency.

Over the last decade, the International Potato Center (CIP) has gained considerable experience in product development and marketing of bakery products in which 20–45% of the wheat flour is replaced with OFSP puree: successfully commercialized in Kenya, Malawi, Rwanda, and Burkina Faso and piloted in Ethiopia, Uganda, Ghana, Nigeria, Tanzania, The Gambia, South Africa, and Mozambique. The OFSP puree value chain activities have created market opportunities for smallholder sweetpotato farmers, especially women and youths, by increasing the value of the crop. Higher farm-gate prices for roots could encourage greater use of farm inputs, which would help to close the existing yield gap, increasing the availability of sweetpotato in rural areas and the accessibility of nutritious OFSP to urban consumers.

5.2 The Evolution of Sweetpotato Puree Processing

The technology for processing OFSP roots into puree was developed in the United States in the 1960s (Kays 1985), e.g., for baby foods, baked products, and other foods. Processing includes root washing, peeling, cutting, steaming/blanching or precooking, finish cooking (with temperature-time programs suitable for starch conversion by endogenous amylolytic enzymes to obtain the products with targeted maltose levels and viscosities), and grinding into puree. In many countries, processing of sweetpotato paste or puree is traditionally practiced at the household and small-scale levels by boiling or steaming fresh roots and then mashing or grinding them. The puree is an ingredient in many foods, including baby foods, soups, pies, cakes, ice creams, breads, and other products (Collins and Walter Jr 1992; Woolfe 1992; Truong 1989, 1994; Truong and Avula 2010).

The development of OFSP beverages, jam, and ketchup with sensory and nutrient attributes comparable to fruit-based products in the early 1990s raised interest among research institutions in several sweetpotato-producing countries including

India, Japan, Malaysia, the Philippines, and the United States (KNAES 1996; Padmaja 2009; Payton et al. 1992; Sankari et al. 2002; Tan et al. 2004; Truong 1992). Transfer of the processing technologies to farmer organizations, small-scale processors, and food companies for pilot testing and market development was conducted in the Philippines (Truong 1992).

In Japan, the puree from white-, orange-, and purple-fleshed sweetpotato varieties was commercially produced by cooperatives and small manufacturers to make ice creams, jams, juices, confectionary, and breads (Duell 1992; Katayama et al. 2017). However, sweetpotato puree was less common than sweetpotato flour as a substitute for wheat flour in bakeries around the world. In two coastal towns of Peru (Cañete and Chincha), OFSP in the form of grated raw or paste of steamed roots (depending on the availability of processing equipment) was substituted for up to 30% of the wheat flour to improve the β-carotene content of a commercial bread, *pan de camote* (Woolfe 1992). Recently, orange-fleshed sweetpotato puree has replaced 20–50% of wheat flour in cookies, donuts, and breads by some commercial bakeries in Ghana, Kenya, Malawi, Rwanda, and Uganda (Bocher et al. 2017).

Challenges for the puree processing industry include the following: (1) adjusting the process to account for differences in carbohydrate content and amylase activities among variety types, postharvest curing, and storage conditions to produce consistent and high puree quality, and (2) preservation technology that produces a shelf-stable product for convenient use in processed foods. Several techniques were later developed to produce purees with consistent quality, despite the varietal differences in carbohydrate content, starch-degrading enzyme activities, and postharvest handling (Collins and Walter Jr 1992).

5.3 Development and Commercialization of Aseptic Shelf-Stable OFSP Puree by Continuous Flow Microwave Processing

Purees can be preserved by canning, freezing, or refrigerated storage with acidification (with citric acid) or by adding preservatives such as potassium sorbate or sodium benzoate (Bocher et al. 2017; Pérez-Díaz et al. 2008; Musyoka et al. 2018a). However, poor product quality due to excessive heat in conventional cooking, the high cost of investment associated with frozen products, and the use of preservatives are the main hurdles for widespread applications of sweetpotato purees in the food industry. An improved canning process was developed for OFSP puree involving flash sterilization and aseptic filling/packaging to produce a shelf-stable product (Smith et al. 1982). However, validating and scaling up the technology did not proceed to commercial development in the United States.

Continuous flow microwave heating and sterilization technologies were initiated at North Carolina State University (NCSU) in the mid-1990s with support from the

Center for Aseptic Processing and Packaging Studies (CAPPS). The initial push for these developments was to enable industrial processing and commercialization of complex multiphase foods in sterile packaging formats, such as low acid soups and stews, and to retain native, typically heat-sensitive flavors, colors, textures, and nutrients like vitamins and antioxidants. Since then, many studies have been performed on other difficult-to-heat and poorly conductive foods and beverages such as salsas, dairy and plant-based milks, smoothies, baby foods, cheese sauces, other sauces, and cooking blends and pureed, homogenized, and diced vegetables and fruits like strawberries, peaches, mangos, blueberries, bananas, and pineapples. To date, over 70 FDA letters of no objections/no questions (LNO) have been issued for a variety of products processed and commercialized using several configurations of continuous flow microwave processing.

In the early 2000s, a novel process for rapid sterilization and aseptic packaging of OFSP puree using a continuous flow microwave system was successfully developed. The product packaged in flexible plastic containers had similar color, carotene retention, and apparent viscosity as the non-sterilized puree and was shelf-stable for at least 12 months (Coronel et al. 2005; Kumar et al. 2008). With this technology, consistently high-quality puree from sweetpotato roots (30–40% of the crop) could be packaged into virtually unlimited container sizes (up to 200 or 250 liters) as a functional ingredient for the food processing industry (Fig. 5.1). In 2008, the developed technology was transferred to a start-up company, Yamco. This large-scale factory has operated with an output of up to 20,000 kg of aseptic puree per day to supply a high quality, shelf-stable functional ingredient to companies manufacturing baby foods, beverages, soups, chips, baked products, and pet foods, among others. The Yamco factory is the first in the world to use a continuous flow microwave sterilization system for the commercial production of shelf-stable OFSP puree.

This rapid expansion of OFSP as an ingredient in food manufacturing has been influenced by the high retained nutritional values and convenient bulk packaging formats for further processing with easy portioning and conveyance/pumping. Previously available puree bulk packaging formats were in cans, 5-gallon plastic buckets of frozen product, or aseptic packaging using conventional (heat exchanger) methods of sterilization, all of which had significant shortcomings in quality levels, energy usage for storage, and ease of use or stability during preparation or defrosting.

After developing and testing the sterilization and shelf-life protocols, the microwave sterilized puree product was compared to similar products produced by other technologies (Table 5.1).

Advanced thermal (microwave) processing of sweetpotato in the U.S. is now used to make dozens of commercial products, including snacks, pie fillings, ice creams, pastas, bakery products, beverages, distilled alcoholic beverages like sweetpotato vodka and bourbon, soups, sauces, vegetable blends, and baby foods.

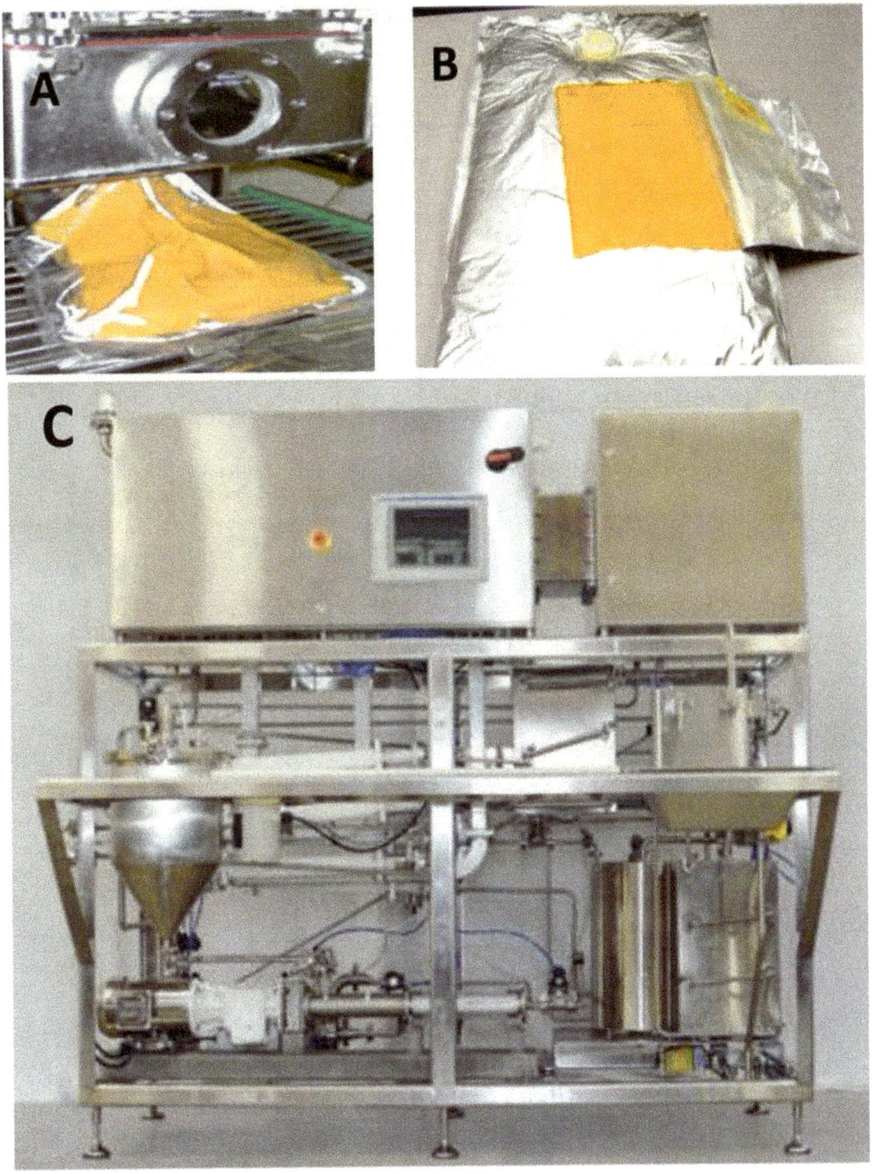

Fig. 5.1 Aseptically packaged OFSP puree produced by the first generation of continuous microwave heating system using a single-head Astepo aseptic filler (**a**) for a pilot scale or LiquiBox StarAsept aseptic filler for commercial production (**b**) in the early 2000s. Photo credits: VD Truong (NCSU). (**c**) Fourth industrial/commercial generation of continuous flow microwave heating/sterilization technology – 24 kW 2450 MHz processing installation at SinnovaTek/FirstWave Innovations in Raleigh, NC. Technology is commercialized by SinnovaTek Inc. under the Nomatic trademark. A factory using this technology has been set up in Kenya since 2019. (Photo courtesy of SinnovaTek Inc.)

Table 5.1 Sweetpotato puree comparisons: continuous flow microwave (first column) aseptic puree vs. other types

	Puree type				
Attributes	Microwave	Fresh	Frozen	Canned	With preservatives
Commercially sterile	Yes	No	No	Yes	Yes
Ambient storage	Yes	No	No	Yes	Yes
Superior natural color	Yes	Yes	No	No	No
Superior flavor	Yes	Yes	No	No	No
Fresh appearance	Yes	Yes	No	No	No
Preferred texture	Yes	No	No	No	No
Ease of use	Yes	No	No	No	Yes
High beta-carotene retention	Yes	Yes	No	No	No
% Preferred (NCSU consumer test)	68	25	1	5	1

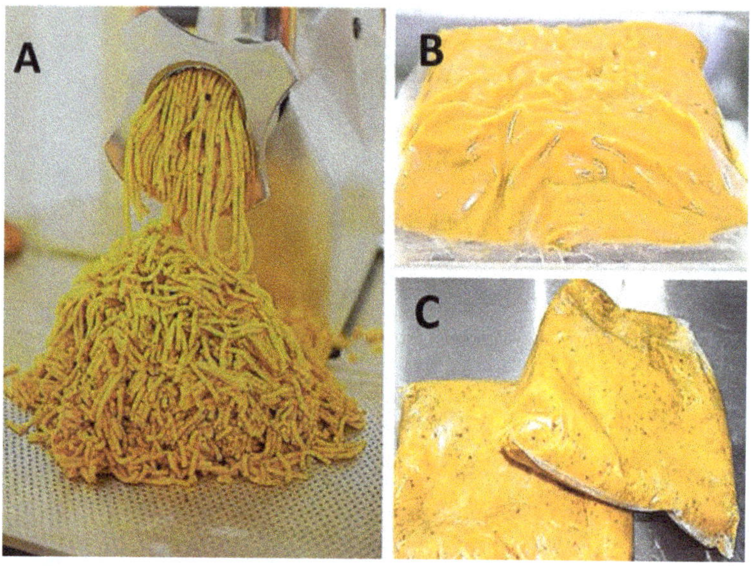

Fig. 5.2 Fresh OFSP puree (**a**) is packed in food-grade plastic bags that are vacuum sealed. (Photo credit: N. Ronoh (CIP)). (**b**) The fresh puree can either be made using peeled roots. (Photo credit: S. Quinn (CIP)). (**c**) A high-fiber puree can be made with unpeeled roots. (Photo credit: CIP)

5.4 The Development of OFSP Puree Inclusive Value Chain in Africa

The rapid growth of African cities is fueling the demand for safe, affordable processed foods, opening up a huge market for OFSP puree as an ingredient in processed foods. Conventionally packaged puree (Fig. 5.2) requires cold chain storage, a major bottleneck to expanding the use of fresh puree in Africa.

Puree processors need to invest in storage facilities to use during the harvest season, to store supplies for the off-season, but this implies a high cost of electricity. Also, the informal food sector is growing in African cities, but few informal traders can afford refrigeration. Therefore, adopting technologies to produce high-quality, shelf-stable OFSP puree in Africa is crucial to allow millions of consumers to access affordable, vitamin A-rich foods throughout the year. This could increase food availability in many food-insecure African countries.

OFSP puree can be treated with preservatives before being packaged and stored for 3–6 months at room temperature. A combination of potassium sorbate, sodium benzoate, and citric acid with vacuum packaging eliminates the need for cold storage for up to 6 months with little loss of β-carotene (Musyoka et al. 2018a). The OFSP puree with preservatives was piloted in Kenya through a partnership with Organi Limited, a sweetpotato processing company based in Homa Bay.

OFSP puree can also be packaged using the aseptic filling technology. This process sterilizes both the product and the packaging material, eliminating the need to add preservatives while prolonging the shelf life of the puree. The OFSP puree can then be stored at room temperature for up to 12 months, enabling informal vendors without cold storage facilities to safely use the product to make cooked foods. A source of affordable packaging pouches was identified, which had been a major bottleneck for scaling this approach. This technology for OFSP puree processing has been piloted in Kenya, Uganda, Ethiopia, and Rwanda through partnerships with private sector entities. It is now being introduced under the BioInnovate project with CIP and partners in Kenya, Rwanda, Ethiopia, and Uganda (BioInnovate 2019). However, further research is needed for this technology to be fully implemented due to the high risk of spore survival and growth of proteolytic *Clostridium botulinum* during storage at room temperatures, with the accompanying high hazard of botulinum toxin formation. Ultimately, a preservative-free OFSP puree is more marketable with more consumers seeking chemical-free food.

As described in Sect. 5.3, OFSP can also be processed into aseptic purees using a continuous flow microwave system to obtain a shelf-stable product. With funding from the UK's DFID, now Foreign, Commonwealth & Development Office (FCDO), under the Development and Delivery of Biofortification at Scale (DDBio) project, CIP partnered with a Kenyan agro-processing company specializing in preserving fruits and other crops, Burton and Bamber (B&B) Limited, to establish the first continuous flow microwave system for sweetpotato purees in Kenya, which will be commissioned in the second half of 2021. CIP worked with researchers from NCSU and SinnovaTek LLC in Raleigh, NC, to provide B&B support with farmer training, certification for Global GAP, and crop management. B&B buys fruits and sweetpotatoes from farmers and sells the produce at supermarkets. B&B encouraged sweetpotato farmers to act as decentralized vine multipliers (DVMs) to supply disease-free sweetpotato planting material (seed) to nearby farmers. Farmers in east and central Kenya were trained in global GAP and to supply OFSP fresh roots for puree processing to B&B.

While investment is increasing in the food sector, small and medium agribusinesses in most developing countries are still capital constrained. New innovative products like biofortified foods require greater visibility and positioning for future commercial investing. CIP's investment in B&B to acquire OFSP puree processing equipment and aseptic packaging is expected to return at least the nominal principal. Positive social and environmental impact is part of the business strategy. CIP also worked with Euro Ingredients Limited (Kenya) to supply the OFSP puree processing technology from Proteo International, Italy. SinnovaTek shipped a microwave processor to Kenya in early 2019. SinnovaTek and Euro Ingredients Limited worked together to install the puree processing and aseptic puree packaging equipment (Fig. 5.1).

CIP and Kenya Agricultural and Livestock Research Organization (KALRO) released new OFSP varieties (Irene and Sumaia) in Kenya that have consumer-preferred appearance and processing traits. Dielectric properties of 12 African genotypes were measured at 915 MHz and 2450 MHz in the temperature range of 22–130 °C. Dielectric properties play a significant role in developing commercial processing systems using microwave energy.

5.5 Scaling and Commercialization of OFSP Puree Processing in Malawi and Kenya

Most wheat flour is imported in Africa: 62% in Kenya, and 99% in Malawi. Scaling and commercializing OFSP puree to make baked and fried foods could replace up to 40% of the wheat flour with a cheaper, nutritious, and locally available product. Bakery products are popular across Africa, where many people eat them for breakfast, lunch, and supper. Bread is widely consumed, with diverse recipes. Its enticing flavor and convenience appeal to consumers, enhancing its popularity (Nwosu et al. 2014).

In Malawi, the Tehilah Bakery and Value Addition Centre has played a major role in scaling OFSP puree. Tehilah is a food processing business, established in 2017, that produces nutritious and healthy foods and adds value to locally grown crops. It was founded by Jean Pankuku, a food technologist, following her practical experience in the food industry, especially in the area of new product development and food processing. Tehilah processes OFSP puree in the conventional form (cold chain) and as baked and fried products. The business currently buys fresh roots from over 1000 OFSP farmers in eight districts (Blantyre, Zomba, Chiradzulu, Mulanje, Thyolo, Balaka, Chikwawa, and Nsanje), increasing economic opportunities not just for sweetpotato growers but also for transporters and workers at the bakery. Tehilah substitutes OFSP puree for 40% of the wheat flour in bread and buns. The bakery in Blantyre sells through over 20 outlets there, including Chichiri Shoprite (Fig. 5.3) as well as to Chikwawa district and Lilongwe, the capital city. The bakery plans to add OFSP cookies and various confectioneries to target new markets such as schools, hotels, offices, and manufacturing companies. Tehilah received recognition from the Ministry of Industry and Agriculture for the successful

Fig. 5.3 Tehilah Bakery substitutes OFSP for 40% of the wheat flour in bread (**a**), sold at Chichiri Shoprite (**b**) and elsewhere in Malawi. (Photo credit: J. Pankuku (Tehilah))

Fig. 5.4 Bread made with OFSP on sale at (**a**) Tuskys Supermarket. (Photo credit: CIP) and (**b**) at Naivas Bakery in Nairobi, Kenya. (Photo credit: C. Bukania (CIP))

commercialization of OFSP puree and notable community impacts by creating jobs for women and youths. Other private companies, such as Mother's Holding Limited, are now also actively involved in OFSP puree processing and product development.

In Kenya, Organi Limited has been actively involved in OFSP puree processing for over 6 years. It is based in Homa Bay County; the cofounder and managing director is Ms. Consolata Bryant. The company has established a network of nearly 200 smallholder farmers of whom at least 50% are women and 20% are youths in the western Kenyan counties of Homa Bay, Migori, Kakamega, Busia, Kericho, and Kisumu. The farmers sell OFSP roots to Organi, which makes about 500 kg of puree daily, with the help of 16 employees. The puree is sold to supermarkets (Fig. 5.4) in Nairobi and Kiambu County. In 2017, the company expanded its product line to include OFSP bread and buns with 40% wheat flour substitution. Organi Ltd is planning to gradually increase the number of fresh root suppliers to at least 1000 by 2025 and to extend the bakery section to industrial capacity by June 2022. This will enable the company to expand the range of OFSP products and to sell to more regions. The Kenyan Ministry of Agriculture and the Ministry of Health have recognized the role of OFSP in improving food security and hidden hunger (e.g., vitamin deficiencies) and have added orange-fleshed sweetpotato to the list of foods that are rich in micronutrients (pro-vitamin A-rich carotenoids). The Ministry of Trade and Industrialization took note of the increasing use of OFSP puree and has now developed specifications to facilitate its use in trade and commercial production. Other private companies, such as Euro Ingredients Limited, are now also actively involved in OFSP puree processing and product development.

The commercialization of OFSP puree in Malawi and Kenya has taken relatively similar paths, with both countries managing to penetrate the bakery industry, with beneficial impacts along the OFSP value chain.

5.6 Consumer Acceptance of OFSP Puree Composite Bread

This section describes consumer acceptability of composite bread from OFSP puree. The focus is on bread because it is widely eaten in Africa (Wanjuu et al. 2019; Atuna et al. 2020). Previous studies show that OFSP puree can be blended with wheat flour to bake consumer-acceptable flat or leavened bread (Table 5.2). The composite bread has a superior aroma, color, and soft texture, which contribute to its acceptability. Other benefits include low production costs and improved vitamin A content (Amagloh 2019). Since favorable sensory assessment scores for OFSP-based bread are one of the key drivers for patronage by consumers (Okello et al. 2021), this product has the potential to be scaled in African communities.

5.7 Food Safety in OFSP Puree Processing in Africa

To maintain consumer trust in OFSP products, the puree must adhere to the highest standard of food safety. Understanding the microflora associated with sweetpotatoes is important if processes are to be designed to control growth of spoilage and pathogenic organisms. While the skin of the root is intact, its flesh is protected from microbial attack. However, once the roots are prepared for puree production, the stages of peeling and cutting disrupt the cell structure, releasing nutrients that can be metabolized by microorganisms (Brackett 1994). Bacteria and fungi make up most of the microflora in low-acid, starchy vegetables (Brackett 1987). How the roots are processed determines the microflora of the puree. First, the OFSP roots are boiled or steamed, killing any vegetative organisms present, but allowing spores to survive. One organism that is associated with many types of foods, including root vegetables, and that produces one of the most potent toxins known is *Clostridium botulinum*, an anaerobic bacterium. Vacuum packaging a puree provides the conditions for the spores to germinate and for this deadly toxin to be produced. Cooking/

Table 5.2 Composite breads including OFSP puree and their consumer assessments

Reference	Flesh color of sweetpotato	Form used	Wheat substitution level (%)	Panelist numbers	Results/deduction
Low and van Jaarsveld (2008)	Orange	Puree Flour	38	98 shoppers in 2004 112 shoppers in 2005	Strong preference for composite bread over wheat-only bread owing to its heavier texture, superior taste, and golden color
Wu et al. (2009)	Not specified	Puree	5, 10, 20, and 30	20 students	Incorporating puree led to a relatively high preference for texture and overall acceptability. The 30% blend was the most preferred
Bonsi et al. (2014)	Orange	Puree	10, 20, and 30	192 consumers	Mean acceptability was 6.25 on a 7-point scale, indicative of high consumer preference for the composite bread
Bonsi et al. (2016)	Orange	Puree	40	50 undergraduates	On a 9-point scale, the overall acceptability for the OFSP-composite bread and wheat-only bread is similar. All other attributes had a score of at least 7, indicating high consumer preference
Awuni et al. (2018)	Orange	Puree	46%	310 consumers	Paired preference assessments indicated that 77% of the respondents preferred the OFSP-composite bread over the existing wheat-only one
Ouro-Gbeleou (2018)	Orange	Puree	Not stated	387 consumers	Scores of respondents showed a preference for OFSP-composite bread based on its sweet taste and soft texture
Wanjuu et al. (2019)	Orange	Puree	30%	1024 consumers	Softness, color, taste, aftertaste, smell, and overall acceptability had scores above 7.4 on a 9-point scale, i.e., "very likable"

steaming is crucial for destroying the vegetative organisms, but post-process contamination can allow microorganisms to reenter the product.

A study in a puree processing plant in Kenya found that steaming roots significantly lowered levels of microorganisms but that levels in OFSP puree increased again due to post-process contamination by organisms such as *Staphylococcus aureus*, a foodborne pathogen that is commonly found on peoples' skin and noses and in infected cuts (Malavi et al. 2018). Fresh puree, which is currently produced in African countries, requires cold storage. However, refrigerated storage (<5 °C) will allow the growth of organisms that thrive at low temperatures, such as *Listeria monocytogenes*, a bacterium that can cause life-threatening illness in vulnerable groups such as pregnant women, young children, the elderly, and people with suppressed immune systems. Preservatives could reduce the chances of contamination. Adding 0.06% (w/v) of the preservative sorbic acid or benzoic acid while reducing the pH of the puree to 4.2 with citric acid inhibited the growth of *L. monocytogenes* at 4 °C (Pérez-Díaz et al. 2008). Acidifying the puree can also control the growth of *C. botulinum* spores that are unable to grow at pH below 4.5 (Lund and Peck 2000). Acidification and preservatives result in an ambient temperature-stable OFSP puree (Musyoka et al. 2018a).

These studies highlight the importance of strict food safety measures in OFSP puree production. In Kenya and Malawi, official quality standards for OFSP puree have been developed and approved by the respective Bureau of Standards, and it will be worthwhile for other African countries to follow suit. Food safety training not only helps change behavior and attitudes of the OFSP puree processors but also contributes to the safety of the products (Malavi et al. 2021). Food safety and hygiene training workshops were conducted in 2017 and 2019 by CIP-Kenya under the RTB scaling project (Musyoka et al. 2018b). Forty-two participants (22 males and 20 females) attended from national and county government, academia, food processing companies, and nonprofit organizations from different counties. The training aimed to equip participants with knowledge and skills on food safety and hygiene. A knowledge-retention evaluation (questionnaire) held in January 2021 found that the trainees had retained 70% of what they learned at the 2019 course. Food safety regulations and trainings are important components for scaling the production and use of OFSP puree (Sect. 5.9).

5.8 Agribusiness Development and Entrepreneurship Opportunities for OFSP Puree in Africa

OFSP varieties with proven agronomic advantages and acceptance by smallholder farmers (Girard et al. 2017) play a great role in the short- and long-term food and nutrition security in sub-Saharan Africa (SSA). From the supply side, improving the productivity of OFSP is hindered mainly by poor access to quality planting material. Farmers usually recycle their seed, so it is often heavily infected with virus diseases and pests (especially weevils). Consequently, yields are typically only 6 tons per hectare, as compared to the potential 21 ton/hectare (Low et al. 2017). To strengthen OFSP seed systems and improve the nutritional status of children under 5 years of age, several new improved biofortified varieties of sweetpotato have been introduced in SSA since 2014 through partnerships between CIP and national agricultural research institutions (NARIs) (Rajendran et al. 2017).

A strong, entrepreneurial sweetpotato seed system has evolved among NARIs to sustain their seed production and provide a consistent supply of quality, disease-free early generation seed (EGS) from community-based decentralized vine multipliers (DVMs). The DVMs are linked to EGS producers (mainly NARIs) who can supply healthy vines (i.e., free of major pests and diseases, such as sweetpotato virus disease (SPVD) and sweetpotato weevil) (Namanda et al. 2011; Rajendran et al. 2017), and they are sold to farmers or supplied through government extension efforts and projects at highly subsidized prices or free of charge (Okello et al. 2015; Ogero et al. 2016; Bentley et al. 2018). Although most sweetpotato farmers still obtain their planting material from their own or neighboring farms, a market for quality vines, particularly for OFSP varieties, has started emerging in African countries such as Kenya, Uganda, and Tanzania (Rajendran et al. 2017). This has created huge opportunities for NARIs, sweetpotato seed and root growers, and also for processors of OFSP products.

5.8.1 Business Models for OFSP Puree Processing

Five potential business models (Table 5.3) for OFSP puree processing have been identified, and the financial viability has been measured. Medium-scale enterprises were the focus of this study because the puree processing equipment is imported and is too expensive for small businesses. The models were identified through a "financial cost-benefit analysis (FCBA)" using real and hypothetical data on technical and financial operations, collected through key informant interviews from industry players in SSA with expertise in this domain (Rajendran et al. 2019). *Model 1* focuses on OFSP puree processors who also make products such as cookies, cupcakes, and the donut-like mandazi. They store the puree briefly without vacuum packaging, and the goods made from OFSP are sold directly to end users. This is an example of the "business to consumers (B2C)" model. *Model 2* focuses on OFSP puree processors who use vacuum packaging and lots of freezers for storage. Puree is then sold to bakeries. This model is categorized as "business to business (B2B)." *Model 3* focuses on bakers and others who buy OFSP puree from processors. *Model 4* focuses on OFSP puree processors who use preservatives and vacuum packaging rather than cold storage. *Model 5* focuses on OFSP puree processors who make use of the hot-fill technology that improves the shelf life of the puree without any need for preservatives.

All five business models are financially viable in the long run, with the level of average investment between $30,000 and $43,000 for medium-scale enterprises. Some key informants said that small enterprises can start with as little as $1000. However, the sensitivity of the business, net present values, internal rate of return, return on investment, and payback period (the period of time required to recoup a capital investment) for the investments differ across the five business models (Table 5.4).

Table 5.3 Types of potential OFSP puree business models and minimum required investment

Business model	Equipment	Final products	Model	Area required (in square meters)	Total annual investment required (USD)	Maximum capacity of puree requirement and annual production (kg)
1	High fiber puree (HFP) machine, steamer, depositor, baking equipment	OFSP puree, OFSP baked and fried products	B2C	36	$43,030	16,000[a]
2	High fiber puree, steamer, vacuum machine with label printer	OFSP puree	B2B	100	$33,370	375,000[b]
3	Depositor (cookie machine), baking equipment	OFSP baked and fried products	B2C	36	$31,390	16,700[a]
4	High fiber puree, steamer, vacuum machine with label printer, ribbon mixer	OFSP puree	B2B	100	$34,170	375,000[b]
5	Hot fill machine, high fiber puree (HFP) machine, steamer, vacuum machine with label printer	OFSP puree	B2B	100	$37,170	375,000[b]

Source: Key informant interviews (KII) with private players in 2018–2019, author's calculation
[a]Investors buy puree from puree producers, which is a requirement for investors per year
[b]Investors produce puree at a maximum capacity per year

Table 5.4 Financial feasibility indicators for potential OFSP puree business models

Business model	NPV	Internal rate of return (IRR)	Discount rate[a]	Payback period (years)	Return on investment (RoI) per year (%)
1	$291,560	85%	18%	1.3	164
2	$247,018	90%	18%	1.4	198
3	$262,342	90%	18%	1.3	203
4	$265,404	77%	18%	2.8	252
5	$345,380	82%	18%	1.0	266

NPV net present value, *IRR* internal rate of return
[a]Discount rate is based on the 18% lending rate in Kenya

The return on investment (RoI) for each business model is more than 100%, and the payback period is a maximum of 3 years (Rajendran et al. 2019). All five models show good financial viability with different intensities probably influenced by different levels of risk factors. It is important for investors to understand the market demand for the products, available funds, and the type of business model that is appropriate for their resource base and setting. There are two significant risks involved in this business: (1) lack of consistent supply of sweetpotato roots and (2) market demand fluctuation for the product. If marketing strategies are well executed, models 1, 3, and 5 can be the most viable ones. If there are fluctuations in the market demand for the puree product, then investors should focus on model 4 where preservatives are used for producing OFSP puree. Though model 2 is financially viable, it comes with the highest risk for investors.

5.9 Scaling Readiness and Strategy for OFSP Puree in Africa

In order to go to scale, the core OFSP puree innovation requires several complementary innovations ranging from availability of planting material and contract farming to cold chain and extension services. With support from RTB, CIP implemented a scaling project between 2019 and 2021 in Kenya, Uganda, and Malawi. The main objective of the project was to increase the use of OFSP puree in popular baked and fried goods. The approach implemented included the following: (1) identifying smallholder OFSP farmers to be trained on good agronomic practices to ensure the production of high-quality roots; (2) product development training for bakers to substitute up to 40% of wheat flour with OFSP puree; (3) creating income and employment opportunities for youths and women along the value chain; and (4) demand creation through increase in awareness of the nutritional and economic benefits of OFSP puree processing and product development. The scaling project worked with small and medium-sized processors to introduce and expand their use of puree and encourage the development of local supply chains providing high-quality OFSP roots.

The scaling project was backstopped by scaling experts from RTB. It used Scaling Readiness, a stepwise approach for analyzing the innovations from a scaling-out perspective, diagnosing the bottlenecks that hinder scaling out, developing scaling strategies, building common agreements with the stakeholders on the strategies, and updating them accordingly (Sartas et al. 2020a). *Scaling Readiness* breaks down the core innovation into components. *Innovation readiness* is the capacity of the innovation to contribute to development outcomes in specific locations. This is presented in nine stages showing progress from an untested idea to a fully mature, proven innovation. *Innovation use* is the level of use of the innovation or innovation package by the project members, partners, and society. Use begins with the team that develops the innovation and broadens to other users who are

Table 5.5 Summary definitions of levels of innovation readiness and use

Stage	Innovation readiness	Innovation use
1	Idea	Intervention team
2	Basic model (testing)	Direct partners (rare)
3	Basic model (proven)	Direct partners (common)
4	Application model (testing)	Secondary partners (rare)
5	Application model (proven)	Secondary partners (common)
6	Application (testing)	Unconnected developers (rare)
7	Application (proven)	Unconnected developers (common)
8	Innovation (testing)	Unconnected users (rare)
9	Innovation (proven)	Unconnected users (common)

Source: Sartas et al. (2020a)

completely unconnected with the team or their partners. Scaling Readiness is a function of innovation readiness and innovation use (Table 5.5).

Readiness and use levels are evaluated using conceptual, applied, and experimental evidence. Scaling Readiness helps to identify the gaps in the design of the innovations and to prioritize the research and engineering work to address these gaps (Sartas et al. 2020b). The Scaling Readiness framework is discussed in more detail in Chap. 3.

Using Scaling Readiness, the team identified ten complementary innovations, their expected outputs, and overall functions (Table 5.6).

5.9.1 *Strategic Partners for Scaling*

In Kenya during 2020, partnerships with actors along the OFSP value chain came into their own during the Covid pandemic. Interactions were more virtual than physical, but because of the previous engagement, there was a positive response that ensured farmers continued to benefit through trade. Farm Concern International helped to strengthen these partnerships and develop new ones to ensure that farmers benefitted. Engagement with informal markets facilitated linkages to farmers. Processors and wholesale buyers were engaged to source OFSP roots from farmers. As a result, farmers were in a position to sell more than 33 metric tons of OFSP roots to various market outlets, both for fresh sweetpotatoes and for puree processing, with an estimated value of 1,205,270 KSh ($11,095). The market share for OFSP in the profiled markets increased as more traders and consumers became aware of the benefits of OFSP. Organi Limited increased its database from 139 to 239 contracted commercial root producers, almost doubling the OFSP puree supply to Naivas supermarket. In Malawi, Tehilah Bakery linked buyers to over 1000 farmers in Nsanje, Balaka, Blantyre, and Thyolo districts. In 2021, there were individual farmers who were selling their roots to the bakery as well as associations. Perisha

Table 5.6 Complementary innovations for scaling of OFSP puree in Kenya, Uganda, and Malawi

Innovation profile – OFSP puree					
Complementary innovation	Output 1	Output 2	Output 3	Value chain level	Innovation function
1. Contract farming to provide disease- and pest-free seed	Large OFSP seed multiplication center for seed production	Quality declared planting material law on commercial vine multiplication	Providing incentives and subsidies to commercial vine producers	Planting material	Accessibility (affordability)
2. Processing-friendly OFSP varieties for production by commercial farmers in Kenya	List of processing-friendly OFSP varieties that can be released quickly in Kenya	List of existing OFSP varieties suitable for different climatic areas		Planting material	Accessibility (availability)
3.Delivery of extension services	Training commercial vine producers on vine multiplication and positive selection	Upgrade extension module on disease- and pest-free OFSP production	Commercial OFSP farmer clusters	Planting material	Accessibility (availability)
4. Cold chain for OFSP puree for sales and shelves for shelf-stable puree	Shelves for storing puree before sales to wholesalers and retailers			OFSP puree	Accessibility (availability)
5. OFSP puree processing technologies, equipment for independent processor	OFSP puree packaging equipment			OFSP puree	Capacity to use (hardware)
6. Demonstrations of OFSP puree processing and packaging equipment	Training workshop on OFSP puree business	Guidelines on procedures of puree making	Temporary holding areas for OFSP fresh roots in the processing facility	OFSP puree	Capacity to use (people)
7. Credit access manual for OFSP producers	Credibility assessment guidelines on OFSP production for banks			OFSP roots	Accessibility (affordability)

(continued)

Table 5.6 (continued)

Innovation profile – OFSP puree					
Complementary innovation	Output 1	Output 2	Output 3	Value chain level	Innovation function
8. Advocacy and awareness campaign on benefits of OFSP	Improved awareness along the sweetpotato value chain			OFSP bread or pastry	Motivation (convince)
9. OFSP puree business development models	Manual for OFSP puree business development			OFSP puree	Capacity to use (people)
10. Climate-controlled storage for OFSP roots	Storage facilities for OFSP to reduce postharvest loses			OFSP roots	Accessibility (availability)

Agro and Packaging Enterprises contracted 140 farmers to supply OFSP roots for puree processing as soon as production starts.

Much attention was given to developing the capacity of small and medium enterprises (SMEs) to run competitive businesses that can access credit from financial institutions. In collaboration with NetBizImpact, the project finalized provision of tailor-made business development services to six SMEs – two from Malawi, three from Uganda, and one from Kenya. The six SMEs were Organi Limited (Kenya), Tehilah Enterprises (Malawi), Perisha Agro and Packaging Enterprises (Malawi), Food Biosciences and Agribusiness Innovation Program (Uganda), EADC (Uganda), and BioFresh Ltd (Uganda). The main goal was to offer business solutions that would facilitate economic potential in the OFSP value chain and make the SMEs involved more competitive in the marketplace and achieve scale. During the coaching and mentorship, the SMEs were helped to identify growth opportunities, key constraints, and support to overcome them. NetBizImpact tailor-made business models aimed at enhancing the SMEs' capacity to take advantage of a promising market potential within the OFSP value chain including the puree. SMEs were encouraged to give their feedback on the relevance of the tools and exercises and effectiveness of the coaching and mentoring sessions included in the business development support. Besides coaching, NetBizImpact helped develop quality profiles for each SME, provided tools and templates, did value chain analysis at firm level, and discussed growth plan development. Discussions were held with NetBizImpact to develop a manuscript on the business development service process.

Naivas Supermarket had a 2-day refresher baking training using OFSP puree to retrain Naivas bakers on how to bake with the puree. The training exercise was held with the food safety training exercise. Some samples were collected for nutritional analysis by CIP's Food and Nutritional Evaluation Laboratory (FANEL), and these showed high levels of pro-vitamin A carotenoids.

A cooking demonstration at the University of Nairobi dietetic kitchen (Department of Food Science, Nutrition and Technology) showed how OFSP puree could be incorporated into foods and contribute to dietary vitamin A. This training for kitchen staff explained the benefits of OFSP puree.

In July, the Cleanshelf Supermarket baking team attended a baking training on using OFSP puree where they were trained to bake with puree to encourage the supermarket to start using OFSP puree.

In partnership with DDBio, BBC Africa, under the Life Clinic segment, broadcasted an episode on OFSP work in Africa, creating awareness of the commercial and nutritional benefits of OFSP for a worldwide audience. Euro Ingredients Limited has continued to provide technology demonstrations, equipment support, and trainings on OFSP product development to different partners in both the formal and informal sectors. Recipes for the incorporation of OFSP puree in commonly consumed Kenyan street foods such chapati, mahamri, and bhajia have been developed. In addition, Euro Ingredients Limited has expanded its supply of the OFSP dough and chapati to also include several informal vendors in middle-income sectors in Nairobi.

In Malawi, Tehilah currently produces a minimum average of 200 OFSP loaves of bread per day to a maximum of 560 loaves a day – depending on the capacity and availability of fresh roots. Production is 6 days a week translating to between 5000 and 13,000 loaves a month. A group of women in Balaka were trained to use OFSP puree to make donuts, fritters, and sweet beer (a locally fermented nonalcoholic drink). Over 10 tons of roots (from farmers associations) were supplied to Mothers Holding Ltd, and 12,480 loaves of OFSP bread were produced by mid-2020. The period also saw the setting up of an OFSP puree processing unit and installation of water and three-phase electricity at Perisha Agro, Malawi.

5.9.2 Core and Complementary Innovations for Scaling

The core and complementary innovations were analyzed using the Scaling Readiness approach. In the three implementation countries, delivery of extension services was the most used complementary innovation, while productive varieties were at the highest level of innovation readiness (Figs. 5.5, 5.6, and 5.7). The next highest level of innovation readiness was for processing-friendly varieties and a credit access manual in Kenya (Fig. 5.5), a business development manual in Uganda (Fig. 5.6), and the advocacy and awareness campaigns in Malawi (Fig. 5.7). Many consumers reject the OFSP varieties because they are low in dry matter, but processors preferred these moister varieties. Processing-friendly varieties were still in development in Uganda and Malawi. Financial constraints limited the implementation of cold chain and climate-controlled root storage. OFSP puree business development

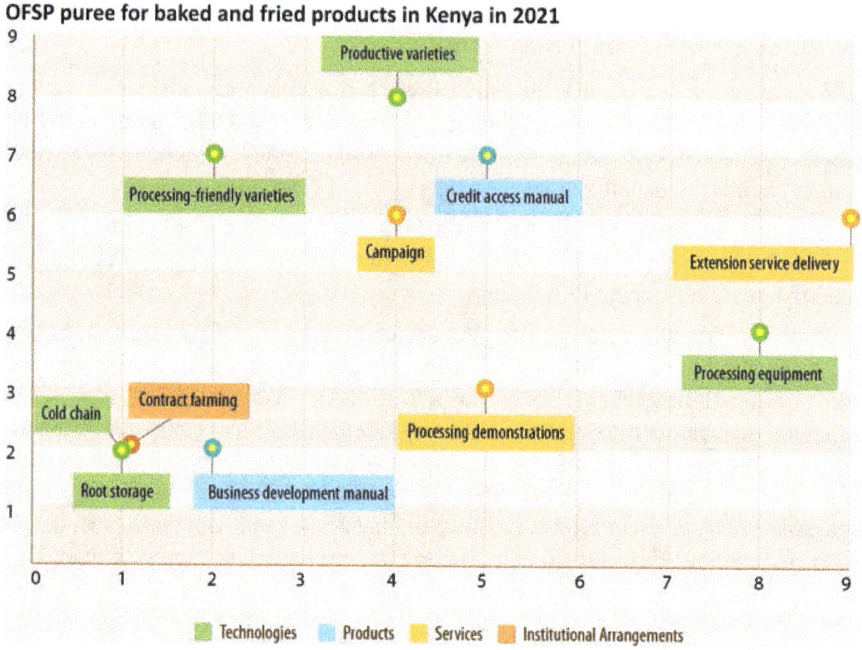

Fig. 5.5 Innovation readiness and use of OFSP puree in Kenya. The *x*-axis represents innovation use, while the *y*-axis represents innovation readiness

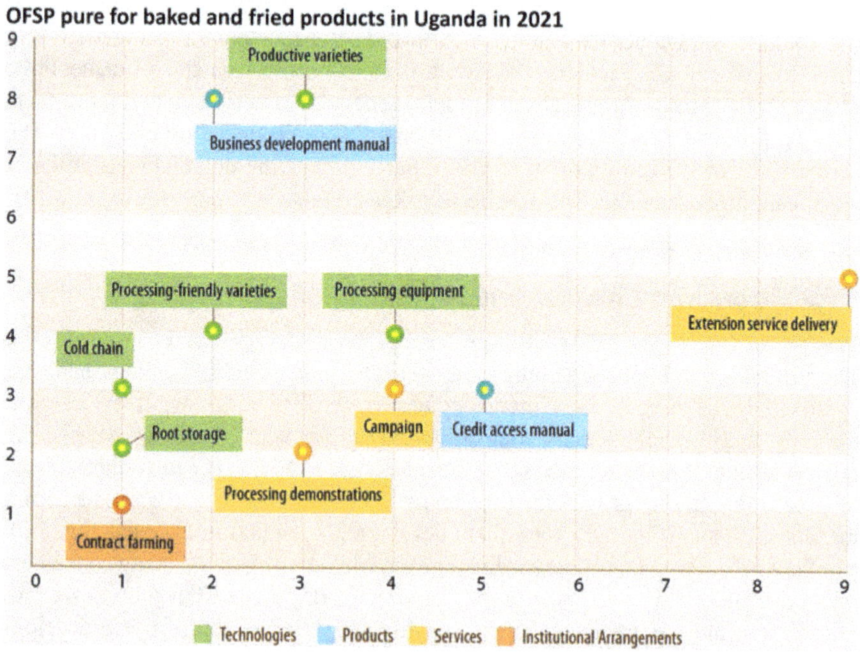

Fig. 5.6 Innovation readiness and use of OFSP puree in Uganda. The *x*-axis represents innovation use, while the *y*-axis represents innovation readiness

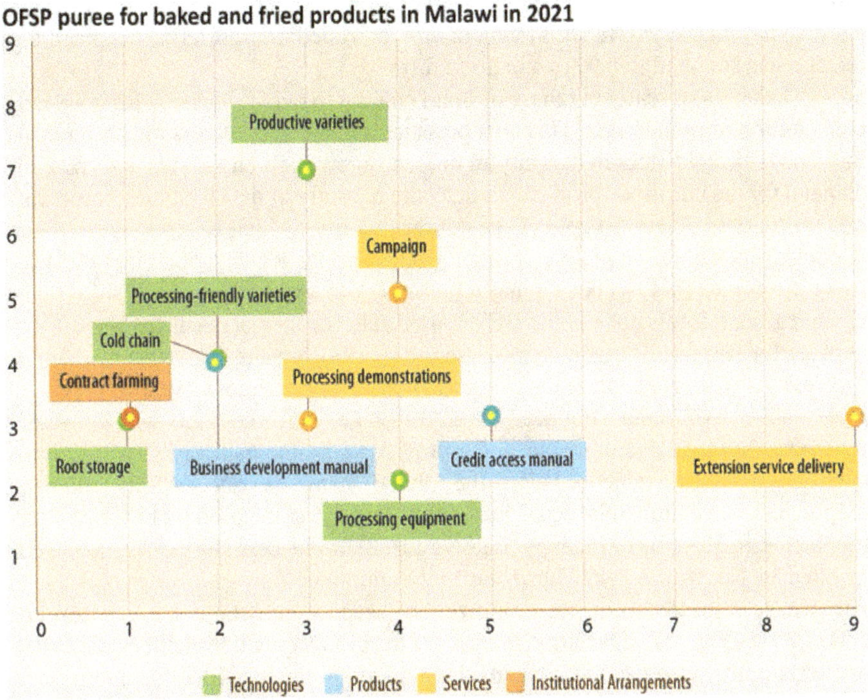

Fig. 5.7 Innovation readiness and use of OFSP puree in Malawi. The *x*-axis represents innovation use, while the *y*-axis represents innovation readiness

manuals were at a high level of innovation readiness in Uganda compared to Kenya and Malawi, while Kenya recorded the highest use of puree processing equipment.

The results highlight some similarities between Scaling Readiness in the three countries and can thus be used as conceptual guidelines for scaling the OFSP puree innovation in other African countries. In addition, they provide guidance on which complementary innovations require more investment to remove bottlenecks.

5.10 Lessons Learned: Moving from Pilot to Large-Scale Commercialization of OFSP Puree in Africa

OFSP puree processing has been successfully piloted in over ten African countries since it was initially introduced with most countries recording steady growth in the demand for OFSP puree-based products. Fresh OFSP puree, which requires cold chain storage, is currently the most common form produced in all African countries. However, the seasonality of the fresh roots poses challenges to the processors. During harvest season, cold storage capacity limits the amount of puree that can be produced and stored in preparation for off-season demand. The need for cold storage also means processors need temperature-controlled transportation to ensure that the puree is supplied to bakeries in good form.

The bakeries typically purchase in small batches to reduce the need for long-term cold storage from their end. This puts pressure on puree processors to continuously increase their cold storage capacity in order to grow their businesses and meet the demand for the nutritious product. To date, the majority of the OFSP processors rely on electricity to power the cold chain storage facilities, which is expensive to maintain in Africa with some regions facing unstable supply, which jeopardizes the quality of the puree. This has been a major bottleneck to scaling the use of OFSP puree in the region. It is this node of the OFSP puree value chain that needs to be addressed in order for large-scale commercialization to become a reality in Africa.

Previous efforts to address the challenges faced by processors included the use of preservatives to eliminate the need for cold storage, which conflicts with consumer preferences. Aseptic shelf-stable puree provides a viable pathway to address this current bottleneck. Yamco LLC in the United States has successfully produced aseptic puree at commercial scale since 2008, supplying the product to various sectors including the bakery industry. Learning from the success story of Yamco, B&B Limited in Kenya has established the first continuous flow microwave system in Africa that is set to commence production of aseptic shelf-stable puree before the end of 2021 (Fig. 5.8). The target market for the product includes large-scale bakeries, restaurants, institutional kitchens, humanitarian organizations for inclusion in nutritional interventions, local governments involved in school feeding programs,

Fig. 5.8 Aseptic puree processing system commercialized by SinnovaTek Inc. under the Nomatic trademark being set up at B&B Limited, Kenya. (Picture credit: Molly Abende, B&B Production Manager)

and informal traders selling street foods, among others. Other countries such as South Africa and Ghana are also in the process of setting up similar systems. Adoption of the continuous flow microwave system in Africa will enable large-scale commercialization of OFSP puree and eliminate the current bottleneck that processors are facing.

5.11 Conclusion

Value addition through processing of agricultural products such as OFSP roots to puree is key to ensuring a stable supply of highly nutritious products to consumers. Puree processing technologies have advanced over the decades from traditional methods involving manual mashing of the cooked roots to highly sophisticated and automated systems such as the fourth generation of continuous flow microwave sterilization systems. To date, steady growth in the commercialization of OFSP puree and the subsequent wheat flour substitution in bakery products has been noted in African countries such as Kenya, Rwanda, Malawi, and Uganda, among others. Positive impacts on revenue generation for small-scale farmers and businesses, employment opportunities for women and youths, and improved nutritional status of target communities are some of the targeted outcomes. These achievements can be attributed to strategic partnerships with stakeholders in agricultural development, public, and private sectors. In addition, financially viable business models have been identified to ensure that OFSP processors make sound financial decisions and maintain a competitive edge in the market. With this rich knowledge base, advanced processing technologies, evidence of Scaling Readiness, and successful piloting of the product, OFSP puree is the breakthrough product for Africa offering the much-needed nutritious products with superior flavor, texture, and appearance accepted by consumers. Further work needs to focus on improving awareness to the multifaceted benefits of OFSP puree along different nodes of the value chain.

Acknowledgments This research was largely undertaken as part of, and funded by, the CGIAR Research Program on Roots, Tubers and Bananas (RTB) and supported by CGIAR Trust Fund contributors. The authors would also like to acknowledge donor agencies including Foreign, Commonwealth & Development Office (FCDO) under the Development and Delivery of Biofortification at Scale (DDBio) project and Swedish International Development Cooperation Agency (Sida) under the BioInnovate Africa Program for the financial support toward the accomplishment of this work. The authors would also like to acknowledge the contributions of private OFSP puree processors in Kenya (Organi Limited, Euro Ingredients Limited, Burton and Bamber Co. Limited), Uganda (Lishe@BIOFRESH), Malawi (Tehilah Bakery and Value Addition Centre), Ethiopia (Duwame Bakery), Rwanda (A Women's Bakery), and Tanzania (AFCO Investment Co. Limited) to the progress reported in this chapter. The authors thank the following individuals who provided input and feedback on earlier drafts of this chapter: Tom Remington, Gordon Prain, and Jeff Bentley. We greatly appreciate Zandra Vasquez's assistance with references and formatting.

References

Abidin PE, Carey E, Mallubhotla S, Sones K (2017) Sweetpotato cropping guide. Africa Soil Health Consortium, Nairobi

Abong GO, Muzhingi T, Okoth MW, Ng'ang'a F, Ochieng' PE, Mbogo DM, Malavi D, Akhwale

M, Ghimire S (2020) Phytochemicals in leaves and roots of selected Kenyan Orange fleshed sweet potato (OFSP) varieties. Hindawi-Int J Food Sci 2020:3567972, 11 p

Amagloh FK (2019) Orange-fleshed sweetpotato composite bread: the nutritional and economic potential. In: Aidoo R, Agbenorhevi JK, Wireko-Manu FD, Wangel A (eds) Roots and tubers in Ghana: overview and selected research papers. KNUST Printing Press, Kumasi, pp 268–283

Atuna RA, Sam FE, Ackah S, Amagloh FK (2020) Bread consumption pattern and the potential of orange-fleshed sweetpotato-composite bread in Ghana. Afr J Food Agric Nutr Dev 20(5):16509–16521

Awuni V, Alhassan MW, Amagloh FK (2018) Orange-fleshed sweet potato (Ipomoea batatas) composite bread as a significant source of dietary vitamin A. Food Sci Nutr 6(1):174–179

Bentley JW, Andrade-Piedra J, Demo P, Dzomeku B, Jacobsen K, Kikulwe E, Kromann P, Kumar LP, McEwan M, Mudege N, Ogero K, Okechukwu R, Orrego R, Ospina O, Sperling L, Walsh S, Thiele G (2018) Understanding root, tuber and banana seed systems and coordination breakdown: a multi-stakeholder framework. J Crop Improv 32(5):599–621. https://doi.org/10.1080/15427528.2018.1476998

BioInnovate (2019) Orange-fleshed sweet potato (OFSP) puree for bakery applications in East Africa. https://bioinnovate-africa.org/orange-fleshed-sweet-potato-ofsp-puree-for-bakery-applications-in-east-africa/

Bocher T, Low JW, Muoki P, Magnaghi A, Muzhingi T (2017) From lab to life: making storable orange-fleshed sweetpotato purée a commercial reality. Open Agric 2(1):148–154

Bonsi E, Chibuzo E, Zibawa R (2014) The preliminary study of the acceptability of Ghana bread made with orange sweet potato puree. J Hum Nutr Food Sci 2(5):1045

Bonsi EA, Zabawa R, Mortley D, Bonsi C, Acheremu K, Amagloh FC, Amagloh FK (2016) Nutrient composition and consumer acceptability of bread made with orange sweet potato puree. Acta Hortic 2016(1128):7–14

Brackett RE (1987) Microbiological consequences of minimally processed fruits and vegetables. J Food Qual 10:195–206

Brackett RE (1994) Microbiological spoilage and pathogens in minimally processed refrigerated fruits and vegetables. In: Wiley RC (ed) Minimally processed refrigerated fruits and vegetables. Chapman and Hall, New York, pp 269–313

Collins JL, Walter WM Jr (1992) Processing and processed products. In: Jones A, Bouwkamp JC (eds) Fifty years of cooperative Sweetpotato research 1939–1989, Southern cooperative series. Bulletin no. 369. Louisiana State University Agricultural Center, Baton Rouge, pp 71–87

Coronel P, Truong VD, Simunovic J, Sandeep KP, Cartwright GD (2005) Aseptic processing of sweetpotato purees using a continuous flow microwave system. J Food Sci 70:531–536

Duell BR (1992) Sweetpotato product innovation by small business in Kawagoe, Japan. In: Hill WA, Bonsi CK, Loretan PA (eds) Sweetpotato technology for the 21st century. Tuskeegee University, Tuskegee, pp 381–389

Girard AW, Grant F, Watkinson M, Okuku HS, Wanjala R, Cole D, Levin C, Low J (2017) Promotion of orange-fleshed sweet potato increased vitamin A intakes and reduced the odds of low retinol-binding protein among postpartum Kenyan women. J Nutr 147(5):955–963. https://doi.org/10.3945/jn.116.236406

Katayama K, Kobayashi A, SakaiT KT, Kai Y (2017) Recent progress in sweetpotato breeding and cultivars for diverse applications in Japan. Breed Sci 67(1):314

Kays SJ (1985) Formulated sweetpotato products. In: Bouwkamp JC (ed) Sweetpotato products: a natural resource for the tropics. CRC Press Inc, Boca Raton, pp 205–218

KNAES (Kyushu National Agricultural Experiment Station) (1996) What's new? Sweetpotato juice. Sweetpotato Breeding. Annual Report for 1996, Number 8. Kyushu National Agricultural Experiment Station, Ministry of Agriculture, Forestry and Foods, Japan

Kumar P, Coronel P, Truong VD, Simunovic J, Swartzel KR, Sandeep KP, Cartwright GD (2008) Overcoming issues associated with the scale-up of a continuous flow microwave system for aseptic processing of vegetable purees. Food Res Int 41(5):454–461

Low JW, Arimond M, Osman N, Cunguara B, Zano F, Tschirley D (2007) A food-based approach introducing orange-fleshed sweet potatoes increased vitamin A intake and serum retinol concentrations in young children in rural Mozambique. J Nutr 137(5):1320–1327. https://doi.org/10.1093/jn/137.5.1320

Low JW, van Jaarsveld PJ (2008) The potential contribution of bread buns fortified with β-carotene-rich sweetpotato in Central Mozambique. Food Nutr Bull 29(2):98–107

Low JW, Mwanga ROM, Andrade M, Carey E, Ball AM (2017) Tackling vitamin A deficiency with biofortified sweetpotato in sub-Saharan Africa. Glob Food Sec 14:23–30. https://doi.org/10.1016/j.gfs.2017.01.004

Low JW, Ortiz R, Vandamme E, Andrade M, Biazin B, Grüneberg WJ (2020) Nutrient-dense orange-fleshed sweetpotato: advances in drought-tolerance breeding and understanding of management practices for sustainable next-generation cropping systems in Sub-Saharan Africa. Front Sustain Food Syst 4:1–22. https://doi.org/10.3389/fsufs.2020.00050

Lund BM, Peck MW (2000) Clostridium botulinum. In: Lund BM, Baird-Parker AC, Gould GW (eds) The microbiological safety and quality of food. Aspen, Gaithersburg, pp 1057–1109

Malavi DN, Muzhingi T, Abong GO (2018) Good manufacturing practices and microbial contamination sources in orange-fleshed sweetpotato purée processing plant in Kenya. Int J Food Sci (2018):11 p. https://doi.org/10.1155/2018/4093161

Malavi DN, Abong GO, Muzhingi T (2021) Effect of food safety training on behavior change of food handlers: a case of orange-fleshed sweetpotato puree processing in Kenya. July 2020. Food Control 119:107500. https://doi.org/10.1016/j.foodcont.2020.107500

Moyo M, Ssali R, Namanda S, Nakitto M, Dery EK, Akansake D, Adjebeng-Danquah J, van Etten J, de Sousa K, Lindqvist-Kreuze H, Carey E, Muzhingi T (2021) Consumer preference testing of boiled sweetpotato using crowdsourced citizen science in Ghana and Uganda. Front Sustain Food Syst. https://doi.org/10.3389/fsufs.2021.620363

Musyoka JN, Abong GO, Mbogo DM, Fuchs R, Low J, Heck S, Muzhingi T (2018a) Effects of acidification and preservatives on microbial growth during storage of orange fleshed sweet potato puree. Int J Food Sci 2018:1–11. https://doi.org/10.1155/2018/8410747

Musyoka J, Njunge F, Muzhingi T (eds) (2018b) Food safety training workshop. Theme: Managing food safety and quality in small-scale food processing for Roots, Tubers and Bananas (RTB) value chains in Sub-Saharan Africa. International Potato Center, Lima, 32 p. isbn:978-92-9060-488-4

Mwanga ROM, Mayanja S, Swanckaert J, Nakitto M, zum Felde T, Grüneberg W, Mudege N, Moyo M, Banda L, Tinyiro SE, Kisakye S, Bamwirire D, Anena B, Bouniol A, Magala DB, Yada B, Carey E, Andrade M, Johanningsmeier SD, Forsythe L, Fliedel G, Muzhingi T (2021) Development of a food product profile for boiled and steamed sweetpotato in Uganda for effective breeding. Int J Food Sci Technol 56:1385–1398. https://doi.org/10.1111/ijfs.14792

Namanda S, Gibson R, Sindi K (2011) Sweetpotato seed systems in Uganda, Tanzania, and Rwanda. J Sustain Agric 35(8):870–884

Nwosu UL, Elochukwu C, Onwurah CO (2014) Physical characteristics and sensory quality of bread produced from wheat/African oil bean flour blends. Afr J Food Sci 8(6):351–355

Ogero K, McEwan M, Pamba N (2016) Chapter 6: Clean vines for smallholder farmers in Tanzania. In: Andrade-Piedra JL, Bentley JW, Almekinders C, Jacobsen K, Walsh S, Thiele G (eds) Case studies of roots, tubers and bananas seed systems, RTB working paper no.2016-3. ISSN 2309-6586., pp 80–97

Okello JJ, Shikuku KM, Sindi K, Low J (2015) Farmers' perceptions of orange-fleshed sweetpotato: do common beliefs about sweetpotato production and consumption really matter? Afr J Food Agric Nutr Dev 15(4):10153–10170

Okello JJ, Shiundu FM, Mwende J, Lagerkvist CJ, Nyikal RA, Muoki P, Mburu J, Low J, Hareau G, Heck S (2021) Quality and psychosocial factors influencing purchase of orange-fleshed sweetpotato bread. Int J Food Sci Technol 56:1432–1446. https://doi.org/10.1111/ijfs.14822

Ouro-Gbeleou T (2018) Boosting demand for biofortified foods: the case of orange fleshed sweet potato bread in Tamale, Ghana. Master's Thesis, University of San Francisco, San Francisco

Owade JO, Abong GO, Okoth MW (2018) Production, utilization and nutritional benefits of orange fleshed sweetpotato (OFSP) puree bread: a review. Curr Res Nutr Food Sci 6(3):644–655

Padmaja G (2009) Uses and nutritional data of sweetpotato. In: Loebenstein G, Thottappilly G (eds) The sweetpotato. Springer Science+Business Media, Germany, pp 189–234

Payton SB, Daniel L, Moore BW, Burnett JA, Stroven JA (1992) Processing method using entire peeled vegetable in a fruit juice/vegetable puree beverage. U. S. Patent 5,248,515

Pérez-Díaz IM, Truong VD, Webber A, McFeeters RF (2008) Effects of preservatives and mild acidification on microbial growth in refrigerated sweetpotato purée. J Food Prot 71:639–642

Rajendran S, Kimenye LN, McEwan M (2017) Strategies for the development of the sweetpotato early generation seed sector in eastern and southern Africa. Open Agric 2(1):236–243. https://doi.org/10.1515/opag-2017-0025

Rajendran S, Magnaghi A, Low J, Muzhingi T (2019) Potential business models and financial feasibility of selected medium-scale business enterprises for orange-fleshed sweetpotato OFSP) value-added products, SASHA brief 22. Sweetpotato Action for Security and Health in Africa Project (SASHA). International Potato Center (CIP), 4 p. https://hdl.handle.net/10568/105951

Sankari A, Thamburaj S, Kannan M (2002) Suitability of certain sweetpotato (*Ipomoea batatas Lam.*) clones for RTS beverage preparation. South Indian Hortic 50:593–597

Sartas M, Schut M, Proietti C, Thiele G, Leeuwis C (2020a) Scaling readiness: science and practice of an approach to enhance impact of research for development. Agric Syst 183:102874. https://doi.org/10.1016/j.agsy.2020.102874

Sartas M, Schut M, van Schagen B, Velasco C, Thiele G, Proietti C, Leeuwis C (2020b) Scaling readiness: concepts, practices, and implementation. CGIAR Research Program on Roots, Tubers and Bananas (RTB). January 2020, pp 217

Smith DA, McCaskey TA, Harris H, Rymal KS (1982) Improved aseptically filled sweetpotato purees. J Food Sci 47:1130–1132, 1142

Tan SL, Omar S, Idris K (2004) Selection of sweetpotato clones with high β- carotene content. In: New directions for a diverse planet: proceedings of the 4th international crop science congress, Brisbane, 26 Sept – 01 Oct 2004

Truong VD (1989) New developments in processing sweetpotato for food. In: Mackay KT, Palomar MK, Sanico RT (eds) Sweetpotato research and development for small farmers SEARCA, Los Baños, pp 213–226

Truong VD (1992) Transfer of sweetpotato processing technologies: some experiences and key factors. In: Scott G, Wiersema S, Ferguson PI (eds) Product development for root and tuber crops. Volume 1 - Asia. Proceedings of the international workshop, April 22–May 1, 1991, VISCA, Philippines, sponsored by CIAT and CIP, Lima, pp 195–206

Truong VD (1994) Development and transfer of processing technologies for fruity food products from sweetpotato. Acta Hortic 380:413–420

Truong VD, Avula RY (2010) Sweetpotato purées and dehydrated forms for functional food ingredients. In: Ray RC, Tomlins KI (eds) Sweetpotatoes: postharvest aspects in food, feed and industry. Nova Science Publishers Inc., New York, pp 117–161

Wanjuu C, Abong G, Mbogo D, Heck S, Low J, Muzhingi T (2018) The physiochemical properties and shelf-life of orange-fleshed sweet potato puree composite bread. Food Sci Nutr 6(6):1555–1563. https://doi.org/10.1002/fsn3.710. PMID: 30258598; PMCID: PMC6145253

Wanjuu C, Bocher T, Low J, Mbogo D, Heck S, Muzhingi T (2019) Consumer knowledge and attitude towards orange-fleshed sweetpotato (OFSP) puree bread in Kenya. Open Agric 4(1):616–622

Woolfe JA (1992) Sweetpotato: an untapped food resource. Cambridge University Press, Cambridge, UK

Wu KL, Sung WC, Yang CH (2009) Characteristics of dough and bread as affected by the incorporation of sweet potato paste in the formulation. J Mar Sci Technol-Taiwan 17(1):13–22

6

Waste Cassava Peels and their Conversion into Animal Feed

Iheanacho Okike, Seerp Wigboldus, Anandan Samireddipalle, Diego Naziri, Akin O. K. Adesehinwa, Victor Attah Adejoh, Tunde Amole, Sunil Bordoloi, and Peter Kulakow

Abstract In Nigeria, processing cassava for food and industry yields around 15 million tons of wet peels annually. These peels are usually dumped near processing centres to rot or dry enough to be burned. Rotting heaps release methane into the air and a stinking effluent that pollutes nearby streams and underground water, while burning produces clouds of acrid smoke. However, when properly dried, peels can be an ingredient in animal feed. Previous attempts over two decades to use peels

---

I. Okike (✉) · P. Kulakow
International Institute of Tropical Agriculture (IITA), Ibadan, Nigeria
e-mail: i.okike@cgiar.org; p.kulakow@cgiar.org

S. Wigboldus
Wageningen University & Research, Wageningen, The Netherlands
e-mail: seerp.wigboldus@wur.nl

A. Samireddipalle
National Institute of Animal Nutrition and Physiology (NIANP), Indian Council of Agricultural Research (ICAR), Bangalore, India

D. Naziri
International Potato Center (CIP), Hanoi, Vietnam
e-mail: d.naziri@cgiar.org

A. O. K. Adesehinwa
Institute of Agricultural Research & Training, Obafemi Awolowo University, Ibadan, Nigeria

V. A. Adejoh
Synergos Nigeria, Maitama Abuja, Nigeria
e-mail: vadejoh@Synergos.org

T. Amole
International Livestock Research Institute, Ibadan, Nigeria
e-mail: t.amole@cgiar.org

S. Bordoloi
Amo Farm Sieberer Hatchery Ltd, Amo Byng Ltd, Awe, Oyo State, Nigeria

in animal feed failed to yield profitable options for drying wet peels at commercial scale, but recent research suggests that cassava peels can be processed into high-quality cassava peel (HQCP) products to be used as nutritious, low-cost animal feed ingredients. The core innovation was to adopt the same steps and equipment used for processing cassava roots into *gari*, the main staple food in the country. When dried, 3 tons of wet peels yield a tonne of healthy and energy-rich animal feed, containing nearly 3,000 kilocalories per kilogram of dry matter (kcal/kgDM). Adopting this innovation at scale in Nigeria's poultry and fish sectors alone has the potential to turn approximately 3.6 million tons of wet peels into 1.2 million tons of feed ingredients capable of replacing approximately 810,000 tons of largely imported maize. The innovation has great potential to increase feed availability and lower its cost while saving cereals for human consumption, reducing the import bill, creating new business opportunities, and protecting the environment. This research was initiated by CGIAR centres and taken up by the CGIAR Research Program on Roots, Tubers and Bananas (RTB) over the past decade with strategic input from the CGIAR Research Program on Livestock to accelerate development of the innovation, and this chapter documents the potential and progress in taking this innovation to scale.

6.1 Introduction

6.1.1 Cassava as an Essential Crop in Nigeria

Cassava (*Manihot esculenta*), also called manioc or tapioca, is a major subsistence and commercial crop in sub-Saharan Africa. Nigeria is the largest producer worldwide, but many of the lessons and opportunities discussed here are applicable to other major cassava-producing countries, particularly in western and central Africa, such as Ghana, Ivory Coast and the Democratic Republic of Congo. Cassava has seen a steady increase in production, averaging 3% per year since 1995, reaching 59 million tons in Nigeria and 178 million tons in Africa (FAOSTAT 2019). Over 90% of the cassava in Africa is consumed as food and very little for industrial processing (Akinpelu et al. 2011). Once considered a subsistence or poor person's crop, cassava is rapidly becoming a commodity with many uses, supporting rural development, poverty alleviation, food security, and value addition with macroeconomic benefits (FAO 2013). Several African governments are promoting cassava to reduce cereal imports through mandatory blending of wheat flour with high-quality cassava flour.

6.1.2 The Growing Demand for Animal Feed and Potential Role of Cassava Peels

Growing population coupled with rising incomes, urbanization and a shift towards protein-rich diets is driving increased demand for meat, dairy and eggs. The demand for animal-derived protein is projected to double by 2050 (FAO 2016), and most of that added demand will come from to low- and middle-income countries (Barry et al. 2012). Between 1999 and 2030, annual meat consumption in these countries will increase from 26 to 37 kg per person, compared with an increase from 88 to 100 kg in industrial countries. Annual per capita consumption of dairy products is expected to rise from 45 to 66 kg in developing countries and from 212 to 221 kg in industrial countries. Egg consumption will grow from 6.5 to 8.9 kg in developing countries and from 13.5 to 13.8 kg in industrial countries (FAO 2011).

The growing demand for animal source food will require more feed (Thornton 2010). Global feed demand is estimated to increase 1.6 times from 1,058 million tons in 2010 to 1,693 million tons by 2050 (OECD-FAO 2013). Growing consensus exists on the tremendous environmental impact of expanding livestock production requiring more land to be brought into production to which the associated increased demand for feed is a major contributor (Willits-Smith et al. 2020). Reducing the pressure on natural resources and ecosystems calls for, among others, a more effective use of available biomass, including recycling (Garnett et al. 2015; Haberl et al. 2014; Karlberg et al. 2015; Smith et al. 2010). Turning crop residues and agricultural waste into feed ingredients is a key strategy to pursue this goal following the circular economy principle.

In Nigeria, the use of cassava roots as animal feed is a traditional backyard practice, but interest from the feed industry for this crop is relatively recent. In 1980, the first international workshop on cassava as livestock feed generated optimism about using cassava roots for replacing cereals, which had become increasingly scarce (Smith 1988). Feed mills in Nigeria were operating at 92% capacity in 1980, but this dropped to 26% in 1997 due to the shortage of local ingredients and the prohibitive cost of imported materials (Azogu et al. 2004), which coincided with an import ban on maize and soybean. Feed mills began searching for substitutes. Cassava chunks, chips and gelatinized grits were the first cassava products to draw attention from feed mills. However, the optimism gradually faded over time as research lagged and cassava was not used in animal feed, largely because yields were stagnant and cassava was growing in demand as food for the growing population.

Livestock can be fed on cassava peels. Farmers and agronomists have known this for years, but the idea had remained largely under-exploited. This situation changed in the 2000s as industry and researchers began to shift their interest from cassava roots to peels and as appropriate technology became available to process wet peels. This development was facilitated by a multi-stakeholder scaling partnership aided by the RTB scaling fund, using the Scaling Readiness approach (Sartas et al. 2020). As a result of this process, high-quality cassava peels (HQCP) have started getting into compound feed for poultry, fish, cattle, sheep and goats in 2018/2019 for some major feed millers, with satisfactory results. We will discuss this process in the following sections.

6.1.3 *The Extent of the 'Peel Problem', Its Underlying Causes and Recent Developments*

In Africa, processing cassava into food is primarily a manual and labour-intensive activity, typically performed by women, either individually at the household level or collectively in specialized processing groups (Hillocks 2002). More than 90% of cassava processing requires hand peeling. As a result, every year, Nigerians generate 15 million tons of peels, stumps (ends of roots) and undersized and damaged roots from processing cassava into various products (Fauquet 2014).

While cassava peels can be fed to livestock, they are poisonous if not properly processed. Peels have high levels of hydrogen cyanide (HCN), particularly in bitter cassava varieties, which are the most common in Nigeria. Sun drying is an effective way to reduce HCN levels significantly. However, improperly dried peels can be contaminated with mycotoxins, toxic compounds produced by fungi that can have serious health consequences for people who consume meat, milk or eggs.

Sun drying takes about 2–3 days in the dry season, but this is not viable in the rainy season when most cassava is harvested, leaving millions of tons of wet peels to rot. In the rainy season, lactic acid-laden, smelly effluent from rotting heaps contaminates nearby streams, wells and underground water (Fig. 6.1). In the dry season, heaps are often set on fire, and the acrid smoke pollutes the air (Fig. 6.2). Millions of tons of fermenting peels also emit greenhouse gases (Obianwa et al. 2016). During the dry season, about half of the processors in Nigeria try to dry up to 20% of their wet peels to sell or feed their own livestock. In rural Nigeria, peels are often given away to the workers who peel the cassava roots, so the peels can be dried and fed to their animals, especially goats (Azogu et al. 2004). However, these practices absorb only a fraction of the massive amount of peels generated in the country annually.

Manual peeling inevitably removes part of the cassava root's flesh along with the peels. It is estimated that approximately 22% of the fresh root weight (including the peels) is removed when roots are manually peeled. Enhancing hand peeling efficiency can reduce this to approximately 18% and save about 270,000 tons of cassava flesh annually, worth US$[1] 17.5 million, in southwest Nigeria alone (Bennett and Naziri 2013). However, procuring these savings would require the adoption of specially designed knives and considerably increase the time required for peeling.

In spite of recent advances in developing cassava peeling machines in Nigeria, it does not appear that the quantity of peels resulting from manual processing

[1] Exchange rate, USD 1 = NGN 480 at the parallel market where the bulk of foreign currency sourcing is happening

Fig. 6.1 A typical dumping site for cassava peels. (Photo: I. Okike (ILRI))

operations will decline in the foreseeable future. Egbeocha et al. (2016) conducted a review of cassava peeling machines and found their performance to be largely unsatisfactory for small-scale processors, who constitute the backbone of cassava processing in Africa. Kolawole et al. (2016) evaluated seven peeling machines and found that peeling efficiency was 0–96% while flesh loss was 7–96%. The latter compares very poorly with the 4% flesh loss for traditional manual peeling and 2–3% for peeling by women who have been specifically trained by the International Institute of Tropical Agriculture (IITA).

However, in spite of their current low efficiency, mechanical peeling is being increasingly adopted by some medium-to-large-scale processing enterprises that value the higher output capacity and the faster processing times. This is particularly true for industrial starch extraction, which usually requires the rapid processing of large amounts of fresh roots and does not need thorough removal of the external cortex that forms the peel with the periderm. For other cassava-based products, female peelers are still used to apply the finishing touches to clean the root of vestiges of peels after mechanical peeling. Therefore, mechanical peeling does not represent an immediate or foreseeable threat to the commercial exploitation of cassava peels nor to women's job opportunities. Unless the efficiency of mechanical peelers is dramatically increased, their wider adoption would simply generate more peels. These peels, in the meantime, could be processed into HQCP mashes (Fig. 6.3).

Fig. 6.2 Attempts to eliminate cassava peels by burning them.

6.2 What Is the Innovation Package?

The core innovation is a processing method for turning wet peels into HQCP mash to use as a feed ingredient. This method circumvents the challenge of drying, which is the most limiting constraint to using cassava peel as feed, reducing drying time from 2–3 days to just 6–8 hours in the dry season by grating, mechanical dewatering, fermenting, pulverizing, sieving and then sun-drying the peels (Fig. 6.4), as detailed by Okike et al. (2015). The steps essentially replicate the age-long processes involved in making cassava storable and safe for human consumption. In the wet season, sun drying is still possible but takes longer due to low intensity of the sun and intermittent periods when the mash spread for drying has to be covered and protected from rain. However, under these conditions, the long time required for reducing the moisture content to the ideal level facilitates fungal growth, which contaminates the product. As such, the sun-drying option is not recommended

Fig. 6.3 An example of peels and wasted roots from a peeling machine, amenable to further processing into HQCP mashes. (Photo: P. Kolawole, IITA)

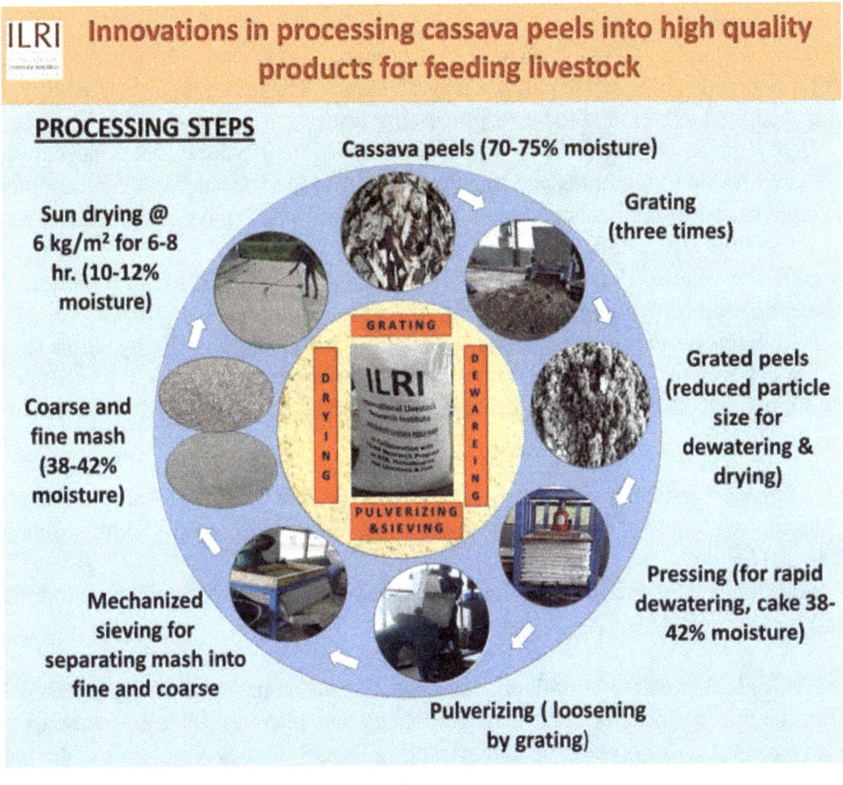

Fig. 6.4 Steps to process fresh cassava peels into high-quality cassava peel products

Table 6.1 Attributes of the four distinct HQCP products

Product type	Attributes	Suitable species to feed
HQCP cake	High moisture (38–42%), low keeping quality (about 7 days). Total energy 2,947 kcal/kgDM and 9.8% fibre	Cattle, goats, sheep and pigs
HQCP whole mash	10–12% moisture and good keeping quality (more than 6 months in storage). Total energy 2,947 kcal/kgDM and 9.8% fibre – Same as for cake	Cattle, goats, sheep and pigs
HQCP fine mash	*Premium product.* 10–12% moisture and good keeping quality (more than 6 months in storage). Total energy 3,039 kcal/kgDM and 8.2% fibre	Poultry, fish and young pigs
HQCP coarse mash	10–12% moisture and good keeping quality (more than 6 months in storage). Total energy 3,039 kcal/kgDM and 15.6% fibre	Cattle, sheep, goats and adult pigs

during the wet season. During wet seasons, additional options for drying include toasting and flash drying.

Regardless of the drying method, four distinct products with different attributes and potential applications result from the process (Table 6.1).

1. *Cassava peel cake* has high moisture content, has a short shelf life of about 1 week and can be fed to cattle, sheep and goats. Due to its high perishability, it is designed to be marketed near processing centres.
2. *HQCP dried whole mash* is drier (moisture content below 12%) and can be stored for up to 6 months and transported over longer distances. It can be a feed ingredient for cattle, sheep, goats and also pigs, which have better capacity to digest fibre than poultry and fish.
3. *HQCP coarse mash* is fibrous and low in energy with a shelf life of 6 months. It can be used for feeding cattle, sheep, goats and adult pigs.
4. *HQCP fine mash* is the premium product, low in fibre, high in energy with a shelf life of 6 months and suitable for poultry, fish and piglets. The HQCP whole mash is mechanically sieved to separate it into HQCP coarse and HQCP fine mashes (Fig. 6.5).

All of these innovative processes reduce the cyanide content to approximately 35 ppm, well below safe levels (90 ppm) for livestock feed (Samireddipalle et al. 2016).

Additional elements of the innovation package include recommendations for feed use, especially:

- FeedCalculator® for formulating least cost balanced rations incorporating HQCP
- Protocols, business development guidelines and plans to guide investments at diverse scales for processing and marketing HQCP mashes (e.g. small scale, two graters with toasters; medium scale, four graters with 4-tonne flash drier related)
- Loan application guidelines for engaging with credit and financial institutions
- District and state-level maps georeferencing cassava processing clusters on Cassava Peel Tracker®, which incorporates information on quantity of peels

Fig. 6.5 A mechanical sieve – coarse mash retained in the wooden box and fine mash captured in the basin after separation.

generated daily by the processing clusters by season and contact details of key persons

- Capacity development for processors on best practices in production, equipment maintenance, personnel safety, product and factory hygiene and cassava peel business registration on Cassava Peel Tracker®
- Bookkeeping

Finally, the innovation package includes a WhatsApp™-based community of practice (CoP) platform known as Cassava Peel First Movers for exchanging knowledge and information and promoting HQCP products. Efforts are also ongoing to have HQCP products recognized by the Standards Organisation of Nigeria (SON) and their standard physical and nutritional qualities listed in the SON manuals.

6.2.1 History of CGIAR Research to Develop the Use of Peels as Feed

In the early 2000s, a flurry of government-backed activities brought together researchers and the public and private sectors to herald a cassava revolution in Nigeria with options for developing a livestock feed industry based on cassava and its derivatives. At that time, the focus was still on the use of cassava roots as feed ingredient. In 2007, the International Livestock Research Institute (ILRI) received a

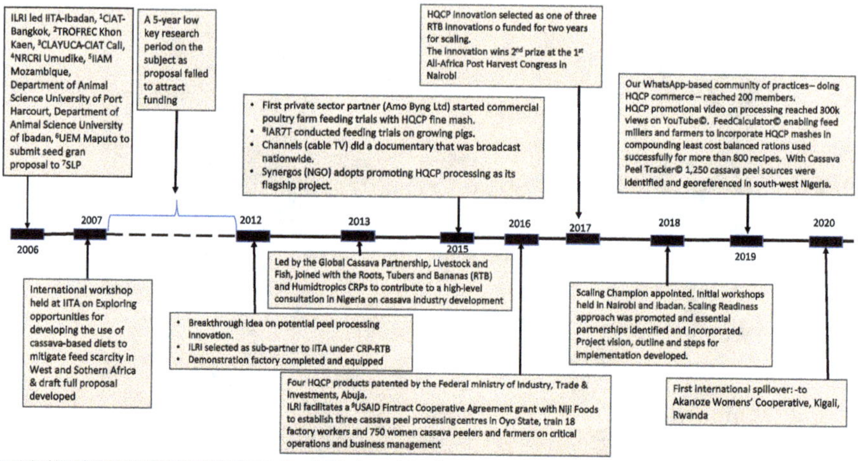

[1]International Center for Tropical Agriculture, [2]Tropical Feed Resources Research and Development Center, Khon Kaen University, Thailand. [3] Latin American and Caribbean Consortium to Support Cassava Research and Development, Cali, Colombia, [4]National Root Crop Research Institute, Umudike, Nigeria, [5]Instituto de Investigacao de Mocambique, Mocambique. [6]Eduardo Mondlane University, Maputo, Mocambique. [7]Systemwide Livestock Project of the Consultative Group on International Agricultural Research (CIGAR).[8]Institute for Agricultural Research and Training Ibadan, Nigeria.[9]United States Agency for International Development.

Fig. 6.6 Timeline of CGIAR research to develop the use of cassava peels as feed

seed grant through the CGIAR Systemwide Livestock Programme (SLP) to conduct initial research on cassava peel as livestock feed. A partnership was later established that included IITA and various research institutes from Nigeria, Mozambique, Colombia and Thailand. The initiative marked the beginning of CGIAR research on the use of cassava peels as animal feed. However, it would not be until the early 2010s that the CGIAR would begin exploring the technical viability of processing cassava peel into HQCP products (Fig. 6.6).

6.2.2 *Proof of Concept and Feeding Trials Under the Auspices of RTB*

After the partnership failed to attract the expected funding, there was a lull in research until 2012, when ILRI engaged in an IITA-led initiative under the CGIAR Research Program on Roots, Tubers and Bananas (RTB). The new initiative aimed at addressing the following research questions:

1. Are HQCP products acceptable to livestock?
2. Are HQCP products economically viable?
3. Are HQCP products have potential for wide adoption and going to scale?

With RTB funding support, a demonstration pilot factory was built, and a multi-disciplinary, multi-institutional team of scientists with expertise in animal science, engineering, food science and pathology came together to investigate options for hygienic and safe processing of wet peels. To complement the RTB effort, the Feed and Forages team of the CGIAR Research Program on Livestock and Fish initiated

Table 6.2 Nutrient profiles and prices of HQCP products compared to maize in March 2021 (dry season)

	HQCP (whole)	HQCP (fine)	HQCP (coarse)	Maize
Starch (%)	66.7	69.0	55.0	68.8
Protein (%)	2.5	2.6	2.8	8.8
Fat (%)	1.4	1.2	1.2	4.1
Crude fibre (%)	9.8	8.2	15.6	2.6
Crude ash (%)	5.8	6.6	3.5	1.4
Total energy (kcal/kgDM)	2947	3039	2495	3840
Price/tonne (naira)	67,000[a]	75,000	50,000	200,000
Price/tonne (US$)	139.56	156.25	104.17	416.67

[a]HQCP products are mainly marketed as fine and coarse mashes although produced as HQCP (whole) before separation. HQCP (whole) yields two parts fine for one part coarse, so its price per tonne can be derived as ((2 × 75,000) + (1 × 50,000))/3 = NGN 67,000 per tonne

a new research activity to explore related small-scale business development services and technology to capitalize on cassava peel waste by-products to increase animal feed availability within selected livestock and fish value chains.

After producing a few tons of HQCP whole, fine and coarse mashes of consistent nutrient profiles and physical qualities (e.g. texture, particle size, moisture level), laboratory analyses confirmed the products were safe with acceptably low levels of cyanide and aflatoxins. The whole HQCP mash was found to have more than two-thirds the metabolic energy (ME) of maize (2,947 vs. 3,840 kcal/kg of dry matter), and these levels were even higher for HQCP fine mash (Table 6.2).

Feeding trials were then established at the ILRI research station with ruminants, poultry and pigs for assessing the palatability of HQCP. Following satisfactory results of the palatability trials and in an effort to create awareness of the products, samples of HQCP coarse mash were taken to ruminant markets across Oyo State, where cattle, sheep and goats consumed it readily.

Most of those involved were from the CGIAR until 2015, when research began to provide solid evidence about the safety and hygiene of HQCP products for animal feed. CGIAR researchers continued research on product development and assessing the technology, but to prove the concept and move towards scaling, partnerships were expanded to include key private-sector players that had toured IITA labs and feeding trials, including ILRI experiments, and expressed interest in setting up their own trials. These groups included Durante Fish Industries, Amo Byng Ltd., Nine Stars Fish Farms and Hatchery Ltd., and Niji Foods. Furthermore, the NGO Synergos and the Federal Ministry of Agriculture and Rural Development (FMARD) were also involved.

On-farm and on-station animal feeding trials were conducted in 2016 at the following three locations.

Table 6.3 Results of feeding trials at Amo Byng

	Groups				
HQCP fed (kg/tonne)	Maize based	50	75	100	125
Chicks housed	100	100	100	100	100
Mean age (days)	35	35	35	35	35
Birds harvested	97	98	97	99	97
Mortality (%)	3	2	3	1	3
Feed consumed (kg)	3.092	3.355	3.047	2.815	2.9
Live body weight (kg)	2.23	2.208	2.184	2.135	2.054
Feed conversion ratio (FCR)	1.387	1.519	1.395	1.319	1.412
Total feed consumed/bird (naira)	265	290	266	247	258
Feed cost/kg liveweight (naira)	118.8	131	121.7	115.6	125.6
Sale value/bird (naira)	892	883	874	854	822
Gross margin/bird (naira)	627	593	608	607	564

Source: Amo Byng Ltd., Awe, Oyo State, Nigeria

1. *Amo Farm Sieberer Hatchery Ltd*, a subsidiary of Amo Byng Ltd., Awe, Oyo State, trialled HQCP coarse mash with feedlot cattle and found it could substitute up to 10% of feed resources. They also conducted four on-farm feeding trials with commercial poultry (three with broilers and one with layers) using HQCP fine fraction. They tested different inclusion rates of mash in the total diet (Table 6.3) and found that the 10% inclusion rate (100 kg of HQCP fine mash in 1 tonne of feed) resulted in the best growth rate, lowest feed intake and lowest mortality. The response in broilers was better than the response in layers. These results convinced the company to start incorporating processed cassava peels into poultry and cattle diets, though supply was a major constraint. Furthermore, HQCP fine mash fortified with full fat soy, DL-methionine and soy oil could replace up to 25% of the maize in broiler diets. In spite of fortified HQCP being significantly cheaper than maize, the biofortification option was not further pursued by the project.
2. *Institute of Agricultural Research and Training (IAR&T) of the Obafemi Awolowo University, Moor Plantation, Ibadan, Oyo State*, conducted on-station trials with weaned and growing pigs. They found that HQCP fine mash could replace 75% of maize in the diet of growing pigs, reducing feed costs by 4% with no adverse effect on the growth performance (Adesehinwa et al. 2016).
3. *ILRI farm at IITA, Ibadan, Oyo State*, hosted on-station trials which showed that cassava peels and leaves, fed in the ratio of 70:30 on a dry matter basis, could be the sole feed for cattle, sheep and goats and would reduce feeding costs.

6.2.3 Supporting Private Sector Partners to Take Up the Innovation

In September 2016, ILRI helped Niji Foods acquire a grant under Fintrac's Cooperative Agreement with the US Agency for International Development

(USAID). Under this initiative, Niji Foods, with support from ILRI, was to establish three HQCP processing centres in Oyo State, train 18 factory employees, six administrative staff and 750 women cassava peelers and farmers on critical operations and business management. Amo Byng Ltd. committed to buying the HQCP products from Niji Foods, but, ultimately, the purchase agreement with Amo Byng Ltd. did not materialize as Niji Foods was offered better prices from other buyers.

6.2.4 Registration of Products as First Step Towards Developing Product Standards

While conducting feeding trials and broadening the partnership, the HQCP project staff approached the Commercial Law Department, Trade Marks, Patent and Designs Registry of the Federal Ministry of Industry, Trade and Investments in Abuja, who granted an umbrella patent to ILRI that covered four HQCP products under trademarks as follows:

1. Cassa peel mash®, previously referred to as HQCP fine mash
2. Cassa peel bran®, previously referred to as HQCP coarse mash
3. Cassa peel cake®, previously referred to as HQCP cake
4. Cassanules®, granulated HQCP that was designed to replace sorghum for the production of Aflasafe®

6.3 Value and Impacts of Innovation

6.3.1 Societal Value of Innovation (Potential Economic and Environmental Value)

The raw material for HQCP production is freely available as approximately 98% of peels in Nigeria are dumped. As previously mentioned, the drying techniques vary according to the season and scale of operations. With sun drying as the dominant mode of processing, the energy requirement is insignificant. Depending on the drying method (sunshine, toasting or flash drying), production costs of HQCP whole mash vary from US$96 to US$107 per tonne (Table 6.4), while the feed industry

Table 6.4 Production budget for HQCP whole mash in the wet and dry seasons (US$/tonne) and market price projections assuming 25% markup during 2015

		A	B	C
Item id.	Item	Budget for fresh peels and initial processing	Budget for a tonne of wet cake = A * 2	Budget for a tonne of whole mash = B * 1.5
a	Fresh peels (dry season)	10.00	20.00	30.00
b	Fresh peels (wet season)	7.00	14.00	21.00
c	Transportation (US$/ tonne/25 km)	1.50	3.00	4.50
d	Loading labour (0.5 h)	0.35	0.70	1.05
e	Off-loading labour (0.5 h)	0.25	0.50	0.75
f	Grating	8.80	17.60	26.40
g	Packing into bags, loading and dewatering by hydraulic press	2.50	5.00	7.50
h	Woven plastic bags (100 reusable bags x 10 kg)	0.50	1.00	1.50
i	**Production of HQCP wet cake and HQCP whole mash (dry season)** = (a + c + d + e + f + g + h) of column B		**47.80**	**71.70**
j	**Production of HQCP wet cake and HQCP whole mash (wet season)** = (b + c + d + e + f + g + h) of column B		**41.80**	**62.70**
k	**Market price of cake assuming 25% markup (dry season)** = i * 1.25		**59.75**	**74.69**
l	**Market price of cake assuming 25% markup (wet season)** = j * 1.25		**52.25**	**65.31**
m	Pulverizing and sieving cake			2.40
n	Labour for spreading and stirring wet mash			6.00
o	Labour for toasting wet mash			10.80
p	Fuel (coal) for toasting			16.80
q	Packing finished products into bags and sealing			4.00
r	Woven plastic bags (40 new bags × 25 kg)			10.00
s	**Production of solar dried mash (dry season)** = (i + m + n + q + r)			**96.10**
t	**Production of toasted mash (wet season)** = (j + m + o + p + q + r)			**106.70**
u	**Market price for solar dried mash assuming 25% markup (dry season)** = s * 1.25			**120.13**
v	**Market price for toasted mash assuming 25% markup (wet season)** = t * 1.25			**133.38**

Source: Analysis based on data from research notes of the cassava peel project

buys HQCP coarse and fine mashes at US$ 100–150 per tonne, respectively, which is less than half the price of maize (Table 6.2). The feed industry is willing to pay for HQCP mashes because of their nutritional content and the satisfactory feedback received from the customers.

Production budgets for both dry and wet seasons indicate that products can reach the market with a markup of 25% on total cost of production and still remain below half the price of maize (weight for weight), which has largely remained the feed industry's benchmark price for HQCP (Table 6.4). Total production cost using sun drying in the dry season is US$97, while the cost of production in the wet season, based on toasting, was US$107 per tonne. A projected market price per tonne of HQCP whole mash of US$120 in the dry season and US$133 in the rainy season would result in a profit margin of 25%, which should keep a business running. However, current market prices (as of Feb 2021) are even higher, approximately US$150 per tonne. It is important to note that while the cost of toasting increases production cost in the rainy season, the price offered for HQCP fine mash is higher during that time because usually, maize prices are highest during the rainy season prior to harvest. Therefore, the additional cost incurred for drying does not necessarily affect profitability adversely.

Cash flow analysis calculated over 5 years of operation for an enterprise producing 2.5 tons of peel cake per day showed that an investment cost of US$9,200 and a working capital of US$100/day would yield a gross revenue of approximately US$39,500 per year, break even in year 4 and result in a net present value (NPV) of US$6,026 (Table 6.5). Other sizes of operations were also analysed with similar results, suggesting a bankable investment (spreadsheets are available on request).

In terms of achievable markets, it is worth noticing that HQCP fine mash is the star product of the core innovation. Because of its lower fibre content and despite

Table 6.5 Five-year budget for an operation processing 2.5 tons of wet cassava peels per day in USD

	Year 0	Year 1	Year 2	Year 3	Year 4	Year 5
Revenue items (USD)						
High-quality cassava peel (wet) cake		37,500	39,375	41,344	43,411	45,581
Gross revenue		37,500	39,375	41,344	43,411	45,581
Cost items (USD)						
Investment (grater, hydraulic press units, well, shed)	9200					
Fresh cassava peels		15,000	15,750	16,538	17,364	18,233
Collation and transportation		1575	1575	1575	1575	1575
Labour and cost of grating		13,200	13,200	13,200	13,200	13,200
Cost of dewatering		5250	5250	5250	5250	5250
Maintenance		200	200	200	200	200
Earnings before taxation		2275	3400	4581	5822	7124
Taxation (1%)		22.75	34	46	58	71
Profit and loss	−9200	2252.25	3366	4535	5763	7053
Present value factor	100%	85%	72%	61%	52%	44%
Present value	−9200	1909	2417	2760	2973	3083
Cumulative present value	−9200	−7291	−4874	−2113	859	3942

Source: Analysis based on data from research notes of the cassava peel project

Internal rate of return = *32%*; working capital, US$35,025 in year 1, going up to US$38,258 in year 5 being roughly 50% more expensive than HQCP coarse mash (Table 6.2), the fine mash is preferred by feed millers for inclusion in feed for monogastrics, especially poultry and fish (incidentally, these are the two most industrialized sub-sectors of the animal production industry in Nigeria and capable of absorbing the highest volume of HQCP fine mash). The potential for HQCP inclusion in the poultry sector is particularly remarkable. Nigeria produced 11.5 million tons of maize and imported 0.5 million tons in 2020. About 60%, or 7.2 million, of these 12 million tons went into poultry feed (Agweek 2021). At a 10% inclusion rate (as suggested by the trials hosted by Amo Farm Sieberer Hatchery), the poultry feed industry alone could replace approximately 720,000 tons of maize per year with same amount of HQCP fine mash. Similarly, if 10% of the approximately 900,000 tons of maize currently used in fish feed manufacturing (CORAF 2020) was replaced by HQCP fine mash, this would create an additional market opportunity of approximately 90,000 tons per year. Therefore, the poultry and fish sectors alone have the capacity to absorb up to 810,000 tons of HQCP fine mash. At a selling price of US$100–150/tonne, this would open up an $80–120 million industry annually that could employ an estimated 20,000 people, 80% of whom would be women.

6.3.2 Who Does This Innovation Package Impact?

Business Opportunities

Wide adoption of HQCP can create a new, commercially viable business and generate income opportunities. Cassava producers, workers, entrepreneurs, livestock farmers and consumers of livestock products can benefit from exploiting this upgrade to the value chain:

1. Approximately four million cassava producers would have the opportunity to sell cassava peels, earning new income that could also spur investment in cassava productivity.
2. *Gari* processors, who generate most of the peels (in some 9000 processing centres nationwide) would enjoy new revenue from selling the waste peels while cleaning up their workplaces (usually nearby their homes), thus improving the health and well-being of workers.
3. Feed millers would have access to cheaper raw materials than maize, even when HQCP is fortified with soy products.
4. Incorporating HQCP into feed mills will reduce feed production costs and release maize for human consumption.
5. Livestock and fish producers would have access to cheaper feed of a similar quality to current products, lowering their operating costs.
6. Consumers would benefit from cheaper meat, milk and eggs, contributing to diets richer in protein and essential micronutrients.
7. Due to increased feed availability, cattle could be produced more intensively with less movement, reducing conflicts between pastoralist and farmers.
8. Dairy production would increase, reducing imports of milk.
9. New job opportunities would be created for people collecting, processing and

selling HQCP products. See also a YouTube video[2] on 'Transforming cassava peels into animal feed'.

Benefits for Women: An Example from Synergos Project

One of the cassava peel project's partners, Synergos, promoted HQCP to women's groups and cooperatives through the State Partnership for Agriculture Programme (SPA), funded by the Bill & Melinda Gates Foundation (2016–2019). The SPA project sought to support women and nutrition in the shift of agriculture from subsistence to commercial production with the goal of increasing incomes and improving livelihoods. Hence, the project provided a good opportunity to Synergos to create a revenue stream for cassava processors, especially women, by establishing prototype HQCP learning centres where cassava peels were collected, processed into HQCP mashes and sold to herders and feed millers.

The SPA used the prototype learning centres to improve relations with farmer cooperatives in Benue and Kogi. In 2016, Synergos funded the training of two women and two men from each of these states as master trainers at ILRI, Ibadan. Ten satellite clusters were equipped with grating machines to process cassava peels and supply HQCP mashes. Over the years, about 2000 women have benefitted directly or indirectly from this innovation in Benue State.

In Kogi State, 20 women and ten female youths were involved in clusters of cooperatives. Synergos, in collaboration with IITA and ILRI, repaired and converted an existing processing centre in Ojapata into a prototype learning centre and helped procure processing equipment. Here, one of the women, Mrs. Jummai Mohammed, aggregates cassava peels from other processing centres in Ojapata and supplies them to the HQCP centre for making fine mash. She sells the fresh cassava peels at 0.52–0.79 per 100 kg bag, and the new income has allowed her to send her children back to school. With scaling fund support from ILRI and the Technology for African Agricultural Transformation (TAAT) initiative, Synergos set up an additional HQCP processing centre for an all-female cooperative group in Ejule. This work was made possible through a partnership in which the women's group provided land, shelter and security for the centre.

In Ogun State, Valueman International also engaged 32 women in an HQCP processing factory, while IFAD and Synergos trained 50 processors, including 30 women. Currently, over 500 women in Nigeria work in the HQCP value chain, and more are projected to benefit as demand increases nationally for HQCP in animal and fish feed production.

The Synergos project has been successful in promoting the HQCP innovation among women (individually and in groups) and impacted positively on the livelihood of beneficiaries. This benefit can be attributed to the coordinated effort of a

---

[2] The video (http://bit.ly/2j7bRu3) has been viewed more than 336,000 times, sparking 2,000 likes and several questions from young school leavers about where to get training and equipment. Clearly, there is interest from young people to engage in such an enterprise.

wide range of partners, including the Ministry of Agriculture and other national agencies (e.g. Business Innovation Facility), research organizations (ILRI and Kogi State University) and private sector actors (Everest Feed and farmers cooperatives).

Though largely successful, there are still some challenges to be addressed and overcome:

1. Access to processing equipment and maintenance. Skills needed to make the equipment are still very low.
2. Transportation to market. It is expensive for HQCP processors to deliver processed HQCP to final markets.
3. Access to credit. Financial institutions are reluctant to loan funds to HQCP processors to invest in equipment.

6.4 The Approach to Scaling the HQCP Innovation

6.4.1 The Influences of the Scaling Readiness Approach

Initially, there was little emphasis on scaling for HQCP. This changed when the lead innovator was named a 'scaling champion' and HQCP transformation was selected as one of the first of three innovations to be supported through the RTB Scaling Fund Project for 2018–2019. The Scaling Readiness approach (Sartas et al. 2020) makes clear that innovations scale as packages, and packages include core and complementary innovations, which, for HQCP, were unpacked into 'hard' (technical) and 'soft' (culture and socioeconomic) components (Fig. 6.7).

Such unpacking of components helps identify what needs to be addressed to help a core innovation go to scale. This work will inform strategic considerations regarding priorities and the roles needed from potential partners by disciplinary advantage. The scaling project team used the concept of 'Scaling Readiness' to assess the innovation packages. 'Innovation readiness' refers to the demonstrated capacity of an innovation to fulfil its contribution to development outcomes in specific locations. This is presented in nine stages showing progress from an untested idea to a fully mature proven innovation. 'Innovation use' indicates the level of use of the innovation or innovation package by the project members, partners and society. This shows progressively broader levels of use beginning with the intervention team who develop the innovation to its widespread use by users who are completely unconnected with the team or their partners. 'Scaling Readiness' of an innovation is a function of innovation readiness and innovation use. Table 6.6 provides summary definitions for each level of readiness and use adapted from Sartas et al. (2020).

Examining components through the Scaling Readiness approach allows researchers and implementers to critically examine the constrained ones while broadening partnership scoping and reflection.

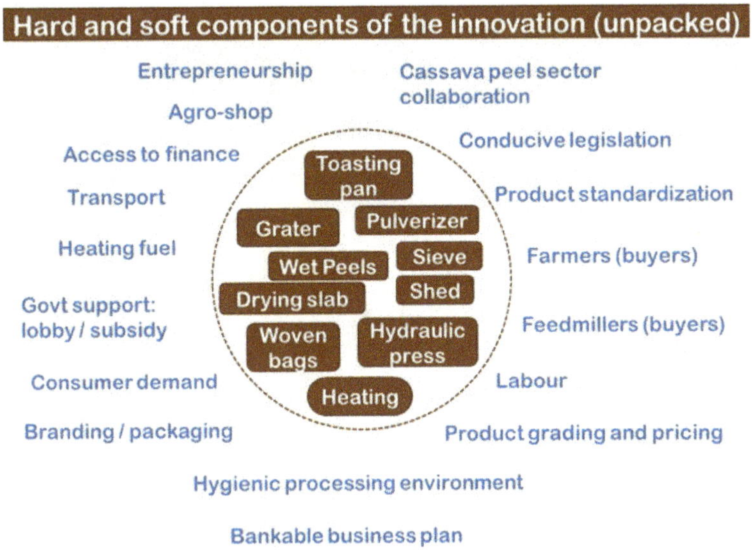

Fig. 6.7 Unpacking the innovation package to reveal its 'hard' and 'soft' components (brown and blue colours, respectively)

Table 6.6 Summary definition of levels of innovation readiness and use (Sartas et al. 2020)

Stage	Innovation readiness	Innovation use
1	Idea	Intervention team
2	Basic model (testing)	Direct partners (rare)
3	Basic model (proven)	Direct partners (common)
4	Application model (testing)	Secondary partners (rare)
5	Application model (proven)	Secondary partners (common)
6	Application (testing)	Unconnected developers (rare)
7	Application (proven)	Unconnected developers (common)
8	Innovation (testing)	Unconnected users (rare)
9	Innovation (proven)	Unconnected users (common)

For HQCP innovation, unpacking and Scaling Readiness analyses revealed that constraints to scaling existed more within the 'soft' components (e.g. environment and product hygiene, processor/user interfacing, grading and pricing, electricity and labour) rather than around the 'hard' components (e.g. grater, hydraulic press and mechanical sieve). If the 'soft' components are the cultural and social economy of the technology, then more partners with competence in these areas are required on board.

In the diagnosis to assess the innovation package in terms of its innovation readiness and innovation use, its individual components were scored on a scale of 1–9 for component readiness and component use. Owing to the large number of components for the innovation package, the diagnosis was partitioned into hard and soft components (Figs. 6.8 and 6.9).

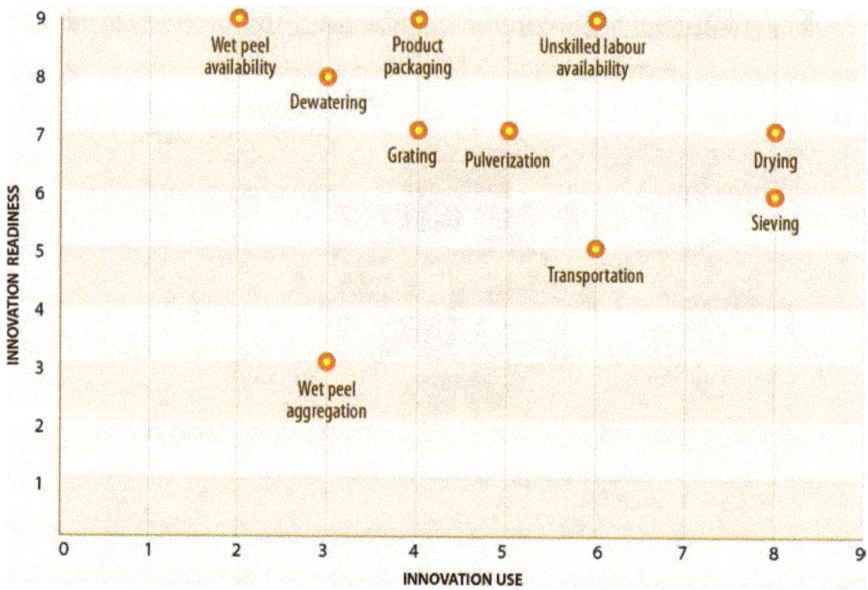

Fig. 6.8 Hard components of the HQCP innovation package. (Source: RTB Scaling Fund Project 2018–2019)

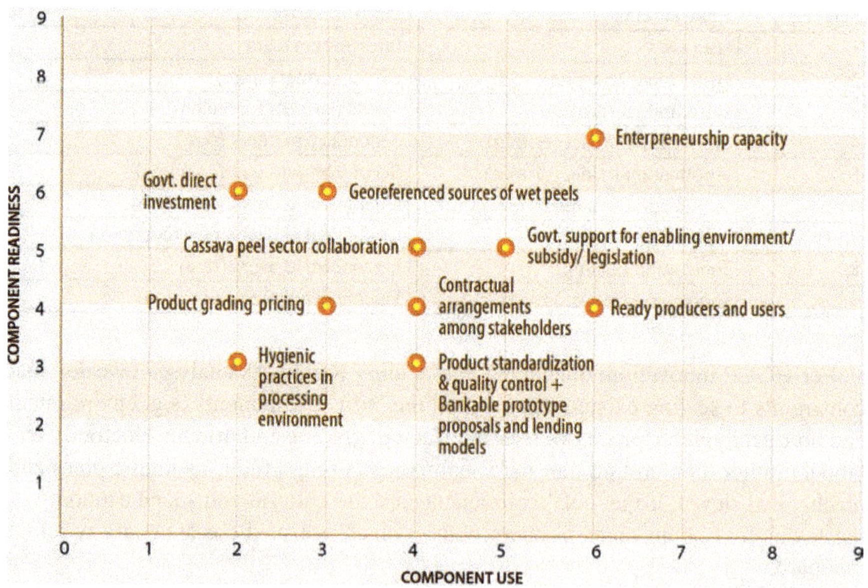

Fig. 6.9 'Soft' components of the HQCP innovation package. (Source: RTB Scaling Fund Project 2018–2019)

From the diagnosis of the hard components, aggregation of wet peels showed the lowest innovation readiness and use, thus constituting an important bottleneck for scaling the HQCP innovation. This triggered the process of developing the scaling strategy considering options to address the identified bottleneck innovations according to Sartas et al. (2020). The initial consideration was to outsource the aggregation of wet peels to a refuse disposal company that would collect and deliver the peels to the processing factories. However, the major problem was the highly dispersed nature of cassava processing centres from which peels have to be collected. Eventually, a web-based platform, called Cassava Peel Tracker®,[3] was developed, and 1,250 locations were surveyed, geo-referenced and mapped in four states in southwest Nigeria to address the bottleneck. The platform provides information on quantity of wet peels generated per location per day (average of 3.3 tons) and the contact details of leaders of the centres. As it turned out, the quantity of peels produced around small enterprises is often sufficient for their operation; however, large enterprises, for whom fresh peels aggregation would be a bottleneck, are yet to emerge.

The diagnosis of the soft components showed that hygiene in the processing environment was at the lowest rung of readiness (Fig. 6.9). The project attempted to tackle this issue at two levels:

(i) At the federal level in the Ministry of Environment, which budgeted to establish six HQCP processing centres in strategic locations to promote hygienic practices. The required funds, however, failed to materialize.
(ii) At the level of individual entrepreneurs through training.

All the options for addressing constraints suggested by the Scaling Readiness approach failed, and as a result, few of the entrepreneurs were able to follow the prescribed hygienic practices (e.g. building underground drainage for effluent). Product standardization and quality control, bankable prototype proposals and lending models were at a similarly low rung of readiness and have remained major constraints to date.

6.4.2 Role of Scaling Workshops in Strengthening Partnerships

The RTB scaling project workshops reiterated the need to focus on the soft components of the HQCP innovation package. This led to the identification of activities needed to address the main bottlenecks for taking the innovation at scale: facilitate access to credit and develop bankable business proposals; establish a cassava peel sub-sector collaboration or community of practice; and ensure product standardization, marketing, promotion, supportive legislation, awareness creation and advocacy.

Based on this, the Bank of Industry (BoI) was engaged to facilitate access to credit, and the National Office for Technology Acquisition and Promotion (NOTAP) began to help with promotion. Single Spark (makers of FeedCalculator® for feed formulation) and the Raw Materials Research and Development Council (RMRDC) began to support the strengthening of the hard components of the technology. Cassava Peel First Movers groups were established as the community of practice. At this point, the scaling consultant helped the project staff develop a way to track partnerships (Table 6.7).

6.5 Development Outcomes from Making and Using HQCP

This section presents the RTB scaling project's achievements and challenges, which were documented through a short, cross-sectional survey conducted shortly after the end of the intervention. The section is structured around the project's development outcomes.

6.5.1 Development Outcome 1

Two major feed millers in Nigeria are regularly incorporating HQCP mash in their commercial rations.

Among major feed millers, Premium Fish Feeds and Premier Feeds incorporate HQCP fine mash directly into their rations, while Nine Stars uses powdered HQCP (indirectly) as a binder in their fish feed. Premium Fish Feeds, which is one of the top five largest feed mills in Nigeria, uses 7–10 tons of HQCP weekly. Premier Feeds began using 100 tons per week and have now expanded to 200 tons a week. Premier Feeds was initially procuring HQCP from their sister factory, Thai Farms, which eventually could not satisfy their expanding demand. Presently, the additional quantity is purchased from one of the project's trainees, Kofo Agro, at a price of US$ 160 per tonne.

6.5.2 Development Outcome 2

Two hundred female and male processors are organized and up to 60% of them linked to markets and selling HQCP mash to commercial feed millers under contractual arrangements.

Small- and medium-scale enterprises that emerged during scaling activities are taking root around HQCP mash production and sale to feed millers.

The post-intervention survey was conducted on three groups of entrepreneurs: (i) 36 of the more than 200 small-scale entrepreneurs belonging to the Cassava Peel

Table 6.7 List of partners and tracking partners

What characterizes partner in relation to scaling project							
Prospective partners	Importance for scaling project	Aware of the innovation proposed and its scaling prospects	Supportive and capable of contributing to scaling	Actively participating in helping innovation go to scale	Actively propagating the innovation without support by scaling	Level of effort to help innovation go to scale	Strategy for how to involve them
	Score 1–5	Tick which one applies				Score 1–5	Description
Amo Byng	4	✓	✓	✓		4	Classification, contractual arrangements for commerce, updates through collaborative fora, participation in meetings
Top feeds	4	✓	✓			1	Classification, contractual arrangements for commerce, updates through collaborative fora, participation in meetings
Durante	3	✓	✓	✓			Updates through collaborative fora, participation in meetings
Nine stars	3	✓	✓	✓	✓	4	Updates through collaborative fora, participation in meetings
Niji foods	3	✓	✓			3	Participation in meetings
Synergos	3	✓	✓	✓	✓	4	Factories and with clientele, updates through collaborative fora, participation in meetings

(continued)

Table 6.7 (continued)

What characterizes partner in relation to scaling project							
Prospective partners	Importance for scaling project	Aware of the innovation proposed and its scaling prospects	Supportive and capable of contributing to scaling	Actively participating in helping innovation go to scale	Actively propagating the innovation without support by scaling	Level of effort to help innovation go to scale	Strategy for how to involve them
NOTAP	4	✓	✓			4	Through collaborative fora, participation in meetings
RMRDC	4	✓	✓			3	Through collaborative fora, participation in meetings
FMARD	4	✓	✓			2	Factories with selected cooperative groups, updates through collaborative fora
BoI	5	✓	✓	✓	✓	3	Beneficiaries for credit support, updates through collaborative fora, participation in meetings
Single spark	3	✓	✓			3	Promotion of FeedCalculator, updates
ILRI	3	✓	✓	✓		2	Project, updates through collaborative fora, participation in meetings
CIAT	3	✓	✓	✓		2	Project, updates through collaborative fora, participation in meetings
Indicative summary	3.5	100%	100%	54%	23%	2.8	

Source: RTB Scaling Fund Project 2018–2019

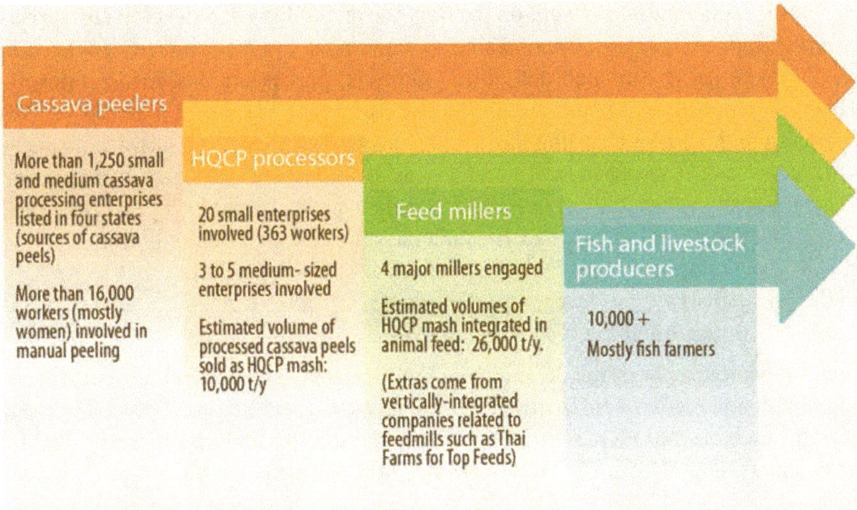

Fig. 6.10 A summary of stakeholders in the HQCP value chain in Nigeria. (Source: End of RTB Scaling Fund project survey, 2020)

First Movers CoP platform, (ii) all three medium-scale entrepreneurs involved with the project and (iii) all four feed millers with whom contractual arrangements were sought. There are 1250 small-scale cassava processors in four states, including more than 200 who belong to the CoP. There are as many as 20 medium-scale entrepreneurs, including those who were not involved with the project (Fig. 6.10).

Small-Scale HQCP Mash Producers

Almost all the 36 small-scale entrepreneurs surveyed had been among the over 200 people trained as a joint RTB-LIVESTOCK CRP activity in the demonstration factory at IITA campus in 2018 and 2019 on best practices in production, equipment maintenance, personnel safety, feed safety, hygiene, cassava peel business registration, bookkeeping, credit application and feed formulation of least-cost balanced rations incorporating HQCP. On a scale of 1 to 5, the training was rated 4.78 for content by respondents. Respondents said the most innovative and useful topics they learned about included how the process mimicked *gari* production, hygiene and feed safety, the use of FeedCalculator® and cash flow analysis.

The respondents already use wet peels as a feed ingredient for their own animals, but now they have identified aggregators for product marketing, good management practices and the promotion of HQCP in feed formulation to reduce costs. For reasons unknown at the time of writing, but probably related to the ease of getting cassava peel, the Cassava Peel Tracker was not popular among small-scale producers and was not used to identify clients or suppliers and/or to buy or sell cassava peels or cassava peel products. Some respondents suggested creating a wider base of processors and buyers to make the website visit worthwhile, while others suggested raising awareness about the existence of the tracker.

Those respondents who went on to adopt the technology and develop enterprises around it (approximately 10% of trainees) were mostly persons already engaged in one or more cassava-related activities – farmers, processors and feed producers. This result is not surprising as the training had a broad waste-to-wealth (w2W) theme around cassava peel that appealed to people already in the cassava business and for whom the required raw material was available free of charge or nearly so. The familiarity of HQCP processing equipment given its similarity to traditional *gari* processing equipment also meant that start-ups or switching from *gari* to HQCP processing was easy for such persons. It is also noteworthy that some 12% of the 36 small-scale producers surveyed started off after watching the demonstration video and/or watching others produce.

As expected, most of the 36 surveyed producers have made fresh investments in equipment and factory construction, while a few switched their previous investment in gari processing to HQCP processing, sensing that it would be more profitable. Investments ranged from US$ 150 to US$ 27,400. More than half of current producers have considered borrowing money to support their businesses, and of these, half of the credit demand is still in preparation. Preferences for credit sources were more for informal and family sources than banks.

A novel industry is developing around the cassava value chain and creating employment in its wake. At the time of the survey, 20 of the 36 respondents were producing HQCP mashes, with enterprises spread across south-east, south-south, southwest and north-central geo-political zones of Nigeria, employing a total of 363 persons (mean 10.8) with weekly production and sales of up to 10 tons of wet peel cake, 14 tons of whole HQCP mash, up to 15 tons of coarse HQCP mash and up to 16 tons of fine HQCP mash (Fig. 6.11).

Medium-Scale HQCP Producers

The three medium-scale HQCP producers who participated in the survey have produced fine mash, coarse mash and cassava peel cake since 2017/2018 for a wide range of clients, including feed millers and fish producers. The medium-scale producers have capacity to process up to 60 tons per week of wet peels and up to 120 tons per week of dried peels. In the past few years, clients have increasingly purchased HQCP products to overcome the problems of traditionally processed cassava peels (dusty and potentially contaminated with HCN and aflatoxins). The customers of the three medium-scale producers rate HQCP as 'most important' on product quality, while availability was the lowest concern. The issue of quality was predicated on nutrient profile (high energy content) and low moisture linked with longer storability and longer shelf life. Longer shelf life helped feed millers even out their HQCP mash supplies and retain better control of their production.

The three medium-scale producers are producing high-quality cassava flour (HQCF), and they source wet peels mainly from their own operations, but they also buy dried peels. The respondents were not sure if they could process more peels than they currently do. An option for expanding operations is to buy wet cake from other producers to process into mashes, but the respondents had no interest in this option being that drying is a major constraint. Nevertheless, these three

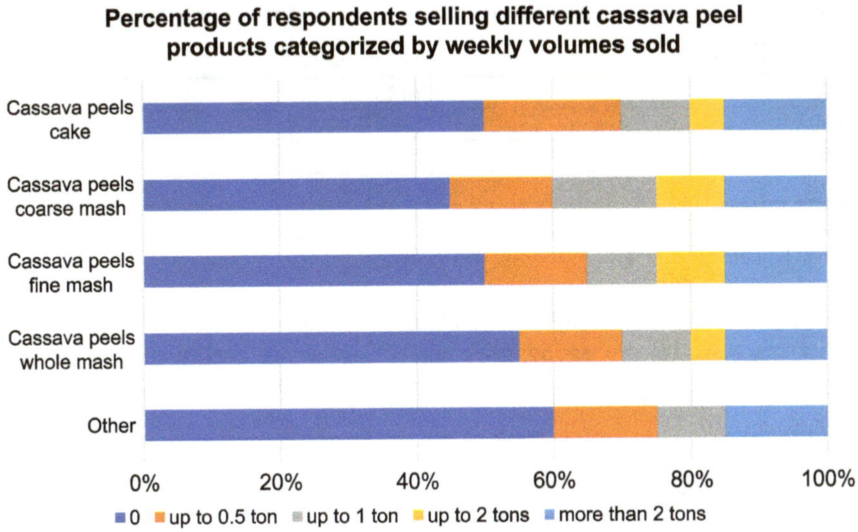

Fig. 6.11 Percentage of the 20 respondents (small-scale processors) selling different cassava peel products categorized by weekly volumes sold. (Source: End of RTB Scaling Fund project survey, 2020)

entrepreneurs plan to expand HQCF production. This development would translate into increased availability of peels, which would expand the HQCP business.

Feed Millers

The four major feed millers who responded to the survey listed several advantages of HQCP, including (1) cheaper than other sources of energy, (2) profitable farmers through reduced feed cost and (3) good source of energy with minimum anti-nutritional factors. They said challenges consisted of (1) low inclusion rate, (2) low crude protein, (3) inconsistent quality and (4) price variation (US$ 125–150). The survey revealed that over 10,000 farmers (mostly fish farmers) have used HQCP products. All of them reported satisfaction with the products, and notably, about one-third of poultry and cattle farmers said they were 'very satisfied'.

Most of feed millers' customers buy poultry and fish feeds. Cattle, small ruminant and other and pig feed clients comprise a small share of the customers (Table 6.8).

Although there was initial reluctance by feed millers, this has been gradually overcome once it became clear that animal production performance remained intact. As such, the incorporation of HQCP in feed has led to increased sales among existing customers and new customers who enjoy lower feed prices. In addition to the 200 tons per week being used by Premier Feeds, up to 520 tons of HQCP is now being incorporated into compound feed by other feed millers every month. This figure is up from 105 tons per month in 2018. Each factory uses different suppliers, but the most reliable supply is through vertical integration (e.g. a cassava flour mill company with a line for HQCP production). All responding feed millers expect to

Table 6.8 Clientele of feed millers that incorporate HQCP mashes by type of feed purchased

Type of feed	No. of clients
Fish	10,132
Poultry	226
Other	21
Cattle	8
Pig	2

Source: End of RTB Scaling Fund project survey, 2020

double their use of HQCP by the end of 2020 barring constraints related to availability, machines for processing, price and consistency in quality and availability. Using more HQCP will lower feed costs and increase factory throughput.

6.5.3 Development Outcome 3

Twenty-five HQCP investors supported to produce successful, bankable proposals and to set up medium to large-scale factories employing youths and women.

The project team developed templates for business plans, including content such as business profiles, vision making and people, business strategy and structure, communication and Information and Communication Technology (ICT) infrastructure, location criteria, operations management strategy, financials and costing, and a marketing and growth program. Different example business plans were prepared for enterprises operating at four different scales. The templates and examples were shared during a 3-day training organized by the project.

Currently, five loan applications are being processed at the Bank of Industry (BoI). BoI staff were part of the trainings and elaborated the requirements for credit. Unfortunately, the gestation period for obtaining credit from BoI, as presented at the training, was overly optimistic. Credit delivery remains encumbered by requirements for securing the loans. However, BoI is not the only financial institution in Nigeria; the business plans are flexible and could be submitted as part of a loan application elsewhere.

6.6 Lessons Learned for Scaling

This section presents the key lessons learned during the implementation of the scaling project. They are categorized in four learning areas.

Capacity/Competencies

The core innovation was developed based on two drying options suitable to small-scale and household enterprises, namely sun-drying and toasting. When the innovation attracted the interest of medium-scale entrepreneurs, additional drying options had to be explored, such as flash drying, rotary drum drying and cabinet drying. The project team lacked the immediate capacities and competencies to respond. Partnership with Niji Foods and collaboration with CIAT experts were critical to overcome these challenges, pointing to the need for careful selection of appropriate partners.

The 'simplicity' of the technology and the decades of processing cassava into *gari* lured us into complacency ('anyone can adopt it') in selecting the initial batches of trainees without sufficient investigation of their motivations. As it turned out, some were interested in the use rather than the production of HQCP, while others were just interested in marketing the products. These are important motivations, but they could not contribute to achieve HQCP production at scale. Based on this experience, interviews were conducted for selecting properly motivated participants to subsequent trainings and to enlist their help to design the training curricula.

HQCP innovation, which builds on age-long processing of cassava into *gari* and other staples, was expected to be a plug-and-play given the similarity of equipment and processes and the huge demand for alternative ingredients in feed production. With an estimated one million households producing and processing cassava into various products and more than 9000 cassava processing centres in Nigeria, we expected immediate massive uptake, but that did not happen. Uptake is never automatic, even when closely related to an existing (technology) way of doing things. However, it is important to note that, when trained in HQCP processing, persons already engaged in some cassava processing had a higher likelihood to become entrepreneurs and adopt the innovation compared to fresh entrants into the cassava processing business.

Strategizing and Scaling Strategy

The high demand for HQCP products could not be met by the small-scale and household producers initially targeted. Meeting this demand will require increasing the numbers of small-scale processors and involve larger-scale investors as well as product aggregators. However, the longer-term aim should be to industrialize the technology so that single factories with capacity to process as much as 50 tons/day become possible and common. There is sufficient demand for HQCP mashes, and fresh peels are available for all scales of producers to the extent that the participation of one group would not adversely affect the other.

Collaboration/Partnership Related

Partnership was clearly biased towards the private sector with a sprinkling of relevant public agencies. The private sector has been involved since the proof-of-concept stage and has stamped a business attitude on the innovation to the extent that it appears that the private sector is leading those big moves.

Early private sector engagement proved helpful to facilitate scaling. For example, with Amo Byng, we performed a technical and economic evaluation of the products and gained their subsequent open endorsement, when bias against HQCP products was still high.

We should do the same and, where necessary, more of reinforcing private sector involvement, engendering transparency and encouraging product standardization to promote commercialization.

Bundling the core innovation with supporting products and services from partners was beneficial to the scaling progress. For example, the collaboration with Single Spark led to the production of FeedCalculator®, which has increased knowledge in feed formulation and promoted the use of HQCP mashes in livestock rations. Practical evidence that inclusion of HQCP mashes into feed did not affect, and sometimes enhanced, livestock performance boosted the confidence of farmers and feed millers and facilitated adoption, especially in poultry and fish production.

The levels of commitment of different partners differed. The partnership with the Bank of Industry, despite being formalized in a contract, did not result in facilitated access to credit by project participants. While formal arrangements can be useful to define roles and responsibilities of partners, these should not be seen as an indicator of commitment to the partnership.

Intervention
Attention to developing the innovation package by adding complementary elements (e.g. hands-on training in production and best business practices, FeedCalculator®, Cassava Peel Tracker® and economic feasibility models) contributed to attract interest to the core innovation – be it for HQCP mash production, feed formulations, pure buying and selling of HQCP products or any combination of the above. These possibilities meant that some of the critical concerns along the value chain regarding new products were addressed within the innovation package.

Overall, it was clear over the course of the entire exercise that scaling had to be a deliberate and proactive effort. *That is, scaling does not just happen, scaling has to be made to happen.* There were several elements (each of them important on their own) that needed to come together, including financial viability and a facilitating business environment. For scaling to happen, each of these elements should be conceptualized not as abstract entities but as having a human face, manifested as interest groups (stakeholders). At this point, one should ask whether these stakeholders desire goals they cannot achieve without certain, mutually beneficial partnerships. If this condition exists, the next step would be to continue to use available, non-costly media to do the groundwork to highlight opportunities created by the technology. Hopefully, this work will arouse different interests and suggest avenues for meeting each other's needs and facilitating agreements that promote commerce. In addition, scaling champions need to follow the positive vibes and remain flexible. There are no blueprints for them to follow.

6.7 Conclusions

Successful scaling requires that the core innovation itself must be easy to use and profitable, like the cassava peel dryers for making HQCP mashes. In this case, the raw materials (cassava peels) were not only free, but they were also an environmental nuisance. At the same time, the rapidly growing livestock industry was demanding more feed. Using HQCP mashes was a chance to turn waste into wealth by using cassava peels to make top quality animal feed. The main challenge was not the hardware (technology), but the software, especially linking small-scale cassava processors (and a few mid-sized ones) with large-scale feed millers. So far, this new value chain is functioning, but it is too soon to tell if the actors will continue to work together.

Doubts remain but these are not insurmountable:

- Will the new product (HQCP mashes) be profitable along the whole value chain?
- Will mutual mistrust remain in spite of the partnerships? Will the lack of formal credit continue to hamper investment?
- Will feed millers continue to buy from small-scale processors (who are mostly women), or will the factories make their own HQCP mashes or buy them from larger processors?
- Will medium-sized producers be able to keep making a quality product, or will they undermine trust in their goods by making them from carelessly dried cassava peels recovered from processor waste piles?

As project funding has ended, it is expected that the work initiated to encourage SON to list HQCP products and set their physical and nutritional standards will provide the guardrails for production of higher-quality products and eliminate poorer quality ones through market forces. Fortunately, the process is still ongoing within SON even in the absence of the project.

The research and scaling efforts have been worthwhile, starting from scratch and gradually nurturing an industry that is in its infancy. Much progress has been made, but support is still needed – especially to buy equipment after training, to standardize products and to promote and market HQCP mashes.

Collaboration with other projects can trigger interest in adopting the HQCP innovation. For instance, the RUNRES project in Rwanda sponsored a training program for a women group in October 2020. During the trip, a skilled fabricator was coached to manufacture the key HQCP equipment. That group now produces about 3 tons weekly based on sun drying even during the rainy season. Another aspect of future work will be to develop HQCP mashes with higher protein content, for instance, through biofortification, which has shown promising preliminary results. This would likely boost the popularity of HQCP mashes among fish and poultry feed millers. The process ferments cassava waste using selected microorganisms, and initial results are positive and indicate the possibility to reach up to 60% conversion rate from waste to feed ingredient on a dry matter basis (Obadina et al. 2006; Aro 2008; Crécy 2012; Tefera et al. 2014).

This innovation has a huge potential to go to scale, since cassava is an important crop in many other countries besides Nigeria. The HQCP project is applicable in all cassava-producing countries. If the importance of cassava in the agricultural sector were to be followed, the Democratic Republic of Congo, Ghana, Cote d'Ivoire, Senegal and Sierra Leone could be possible candidates to scale to innovation. However, the needs of their feed industry should be carefully investigated.

Acknowledgements This research was largely undertaken as part of, and funded by, the CGIAR Research Program on Roots, Tubers and Bananas (RTB) with contributions from the CGIAR Research Program on Livestock, its predecessor, the CGIAR Research Program on Livestock and Fish and the CGIAR Research Program on Integrated Systems for the Humid Tropics and supported by *CGIAR Trust Fund contributors* and initial core funding from the International Livestock Research Institute (ILRI) and the CGIAR Systemwide Livestock Programme.

References

Adesehinwa AOK, Samireddypalle A, Fatufe AA, Ajayi E, Boladuro B, Okike I (2016) High quality cassava peel fine mash as energy source for growing pigs: effect on growth performance, cost of production and blood parameters. Livest Res Rural Dev 28(11)

Agweek (2021) Market output report: Nigeria maize market outlook 2021/22. Agweek 1-1, January 2021

Akinpelu AO, Amamigbo LEF, Olojede AO, Oyekale AS (2011) Health implications of cassava production and consumption. J Agric Social Res (JASR) 11(1):118–125

Aro SO (2008) Improvement in the nutritive quality of cassava and its by-products through microbial fermentation. Afr J Biotechnol 7(25):4789–4797

Azogu I, Tewe O, Ezedinma C, Olomo V (2004) Cassava utilization in domestic feed market in Nigeria. National Centre for Agricultural Mechanization (NCAM), Federal Ministry of Agriculture and Rural Development (FMARD), Abuja, November 2004, 156 pp

Barry MP, Linda SA, Shu WN (2012) Global nutrition transition and the pandemic of obesity in developing countries. Nutr Rev 70(1):3–21. https://doi.org/10.1111/j.1753-4887.2011.00456.x

Bennett B, Naziri D (2013). Market Study for a range of potential cassava and yam waste product solutions in Ghana, Nigeria, Thailand and Vietnam. https://www.researchgate.net/publication/267096062_Market_Study_for_a_range_of_potential_cassava_and_yam_waste_product_solutions_in_Ghana_Nigeria_Thailand_and_Vietnam

Crécy EF (2012) Experimental evolution of a facultative thermophile from a mesophilic ancestor. Appl Environ Microbiol 78(1):144–155

CORAF (2020) Integrated regional strategy for sustainable management of agri-inputs in West Africa and the Sahel. West and Central Africa Council for Agricultural Research Development (CORAF). Dakar. 2020. 84 pp

Egbeocha CC, Asoegwu SN, Okereke NA (2016) A review on performance of cassava peeling machines in Nigeria. Futo Journal Series (FUTOJNLS) 2(1):140–168

FAO (2011) Mapping supply and demand for animal-source foods to 2030, by T.P. Robinson & F. Pozzi. Animal Production and Health Working Paper. No. 2. FAO Rome

FAO (2013) Save and Grow Cassava- A guide to sustainable production intensification. ISBN 978-92-5-107641-5. FAO Rome

FAO (2016) Meat and meat products. FAO's Animal Production and Health Division. http://www.fao.org/ag/againfo/themes/en/meat/home.html

FAOSTAT (2019) Food and Agriculture Organization of the United Nations, 2019. Production: Crops. http://faostat.fao.org.

Fauquet MC (2014) Estimating cassava wastes in Nigeria: a consultancy report submitted to the International Livestock Research Institute (ILRI) Nairobi

Garnett T, Roos E, Little DL (2015) Lean, Green, Mean, obscene...? What is efficiency? and is it sustainable? Animal production and consumption reconsidered Technical Report. Food Climate Research Network, University of Oxford, 48 pp. http://hdl.handle.net/1893/24127

Haberl H, Erb K-H, Krausmann F (2014) Human appropriation of net primary production: patterns, trends, and planetary boundaries. Annu Rev Environ Resour 39:363–391. https://doi.org/10.1146/annurev-environ-121912-094620

Hillocks RJ (2002) Cassava in Africa. In: Hillocks RJ, Thresh JM, Bellotti A (eds) Cassava: biology, production and utilization. CABI Publishing, New York, pp 41–54

Karlberg L, Hoff H, Flores-López F, Goetz A, Matuschke I (2015) Tackling biomass scarcity—from vicious to virtuous cycles in sub-Saharan Africa. Curr Opin Environ Sustain 15:1–8. https://doi.org/10.1016/j.cosust.2015.07.011

Kolawole P, Samuel T, Alenkhe B, Diallo T, Abass A, Kulakow P (2016) Evaluation of commercial cassava peeling machines in Nigeria. Unpublished research manuscript. International Institute of Tropical Agriculture (IITA) Ibadan

Obadina AO, Oyewole OB, Sanni LO, Abiola SS (2006) Fungal enrichment of cassava peels proteins. Afr J Biotechnol 5(3):302–304. http://www.academicjournals.org/AJB

Obianwa C, Uyoh EA, Igile G (2016) Bioethanol production from cassava peels using different microbial inoculants. Afr J Biotechnol 15(30):1608–1612. https://doi.org/10.5897/AJB2016.15391

Okike I, Samireddypalle A, Kaptoge L, Fauquet C, Atehnkeng J, Bandyopadhyay R, Kulakow P, Duncan AJ, Alabi T, Blummel M (2015) Technical innovations for small-scale producers and households to process wet cassava peels into high quality animal feed ingredients and aflasafe™ substrate. Food Chain 5(1–2):71–90. https://doi.org/10.3362/2046-1887.2015.005

Samireddipalle A, Bolakale O, Fabamise A, Okike I, Blümmel M (2016) Performance of West African Dwarf rams fed cassava by products as sole feed. ANACON-2016.X Biennial Conference of Animal Nutrition Association. 09–11 November 2016; College of Veterinary Sciences, SVVU, Tirupati

Sartas M, Schut M, Proietti C, Thiele G, Leeuwis C (2020) Scaling readiness: science and practice of an approach to enhance the impact of research for development. Agric Syst 183(102874). https://doi.org/10.1016/j.agsy.2020.102874

Smith OB (1988) A review of ruminant responses to cassava-based diets. In: Hahn SK, Len R, Egbunike GN (eds) Cassava as livestock feed in Africa: proceedings of the IITA/ILCA/University of Ibadan workshop on the potential utilization of cassava as livestock feed in Africa, 14–18 November 1988, Ibadan, pp 39–53

Smith P, Gregory PJ, Van Vuuren D, Obersteiner M, Havlík P, Rounsevell M, Woods J, Stehfest E, Bellarby J (2010) Competition for land. Philos Trans R Soc B 365(1554):2941–2957. https://doi.org/10.1098/rstb.2010.0127

Tefera T, Ameha K, Biruhtesfa A (2014) Cassava based foods: microbial fermentation by single starter culture towards cyanide reduction, protein enhancement and palatability. Int Food Res J 21(5):1751–1756

Thornton PK (2010) Livestock production: recent trends, future prospects. Philos Trans R Soc. B 365(1554):2853–2867. https://doi.org/10.1098/rstb.2010.0134

Willits-Smith A, Aranda R, Heller MC, Rose D (2020) Addressing the carbon footprint, healthfulness, and costs of self-selected diets in the USA: a population-based cross-sectional study. Lancet Planet Health 4(3):e98–e106. https://doi.org/10.1016/S2542-5196(20)30055-3

7

Cassava Processing: Technology Transfer Activities

Alfredo Augusto Cunha Alves, Luciana Alves de Oliveira, and Joselito da Silva Motta

Abstract Cassava is currently the fourth most important food production crop in tropical and developing countries. Cassava root and its by-products are the main source of calories for the diets of 800 million people in Africa, South America, and Southeast Asia. Over the past 20 years, the Brazilian Agricultural Research Corporation (Embrapa) and collaborators have been developing innovations for the use and postharvest processing of cassava. These technologies have been transferred and disseminated to technicians, entrepreneurs, producers, and processors of cassava from several African countries. This South-South cooperation has been conducted in Brazil through short trainings, workshops, and technical visits requested by national R&D institutions, cooperatives, cassava producers, and processors associations and sponsored by international agencies and foundations. In this chapter, we present an overview of the technology transfer activities of *Embrapa Mandioca e Fruticultura* carried out for Africa, focusing on technological innovations that result in products and by-products of cassava root processing, especially those with great potential for adoption and opening new markets for Africa (e.g., precooked and frozen cassava, cassava chips, among others). The selection of these innovations was based on observations of the trainees' preferences and interests for technologies that they envisioned willingness to apply and share the technology when returning to their countries.

7.1 Introduction

Cassava is one of the oldest cultivated crops in the world; its domestication occurred approximately 9000 years ago in the Amazon region of Brazil (Alves-Pereira et al. 2018). In tropical and developing countries, cassava holds great economic and nutritional importance and feeds around 800 million people. Of the 278 million tons of cassava roots produced by 101 countries in 2018, about 60% is grown on the African

A. A. C. Alves (✉) · L. A. de Oliveira · J. S. da Silva Motta
Embrapa Mandioca e Fruticultura, Cruz das Almas, Brazil
e-mail: alfredo.alves@embrapa.br; luciana.oliveira@embrapa.br; joselito.motta@embrapa.br

continent, 30% in Asia, and 10% in the Americas (FAOSTAT 2019). Cassava is the fourth most important source of calories for the human diet in Africa, South America, and Southeast Asia, behind wheat, rice, and maize. Five African countries are among the ten largest cassava-producing countries, which account for 75% of world production (FAOSTAT 2018).

Cassava cultivars are classified in two types in Brazil: *sweet cassava*, aka "*aipim*," "*macaxeira*," or "table cassava," and *bitter cassava*, aka "*mandioca brava*" or "cassava for industry." These types are classified based on levels of cyanide (HCN) release, a highly toxic substance if ingested. Sweet cassava contains less than 100 mg kg^{-1} of cyanogenic compounds (CC) per fresh root, while bitter cassava contains more than 100 mg (Araujo et al. 2019). High levels of CC require a complete detoxification process to reduce compounds to a safe level for consumption (Montagnac et al. 2009). Cassava roots are very perishable and have a short postharvest shelf life, which severely limits their potential in the market and potential benefits to farmers. The roots exhibit visible symptoms of postharvest physiological deterioration (PPD) within only 24–72 hours of harvest (Morante et al. 2010). The darkening caused by physiological deterioration is an important factor that should be considered in root processing. Like other root and tuber crops, cassava has high water content (~65%), which is probably the major limitation to improving the utilization potential of the crop, requiring rapid processing into intermediate products to reduce transport costs, increase shelf life, and improve storage capacity (Falade and Akingbala 2011).

In Brazil, both the cassava root and the starch extracted are processed and used to make a series of products that can be consumed directly or used as ingredients in a wide variety of foods (Fig. 7.1) (Oliveira et al. 2020a; Tomlins and Bennett 2017).

In the last 20 years, the Brazilian Agricultural Research Corporation (Embrapa) has received several demands from African cassava-producing countries and funding agencies for development projects, requesting technical collaboration from Brazil to transfer technological innovations related to the cultivation and postharvest processing of cassava. Cassava cultivation in Brazil and many parts of Africa shares many similarities between Brazil and Africa in both cultural and social aspects. The profile of cassava producers is similar, and most cassava is produced on family farms. In addition, the agroecological similarities and the great diversity of cassava processing products available in Brazil facilitate the transfer and adoption of these technologies to Africa.

Embrapa Mandioca e Fruticultura international technical cooperation program started in 2000 with the Brazil-Japan Partnership Program as a means to offer courses to technicians from Portuguese-speaking African countries (e.g., Angola, Cape Verde, Guinea-Bissau, Mozambique, São Tomé and Príncipe) and Timor East in Southeast Asia. The program was financed by the Japan International Cooperation Agency (JICA).

In this program, courses were offered to technicians, researchers, and representatives of national agricultural research institutions (NARIs), and the content focused

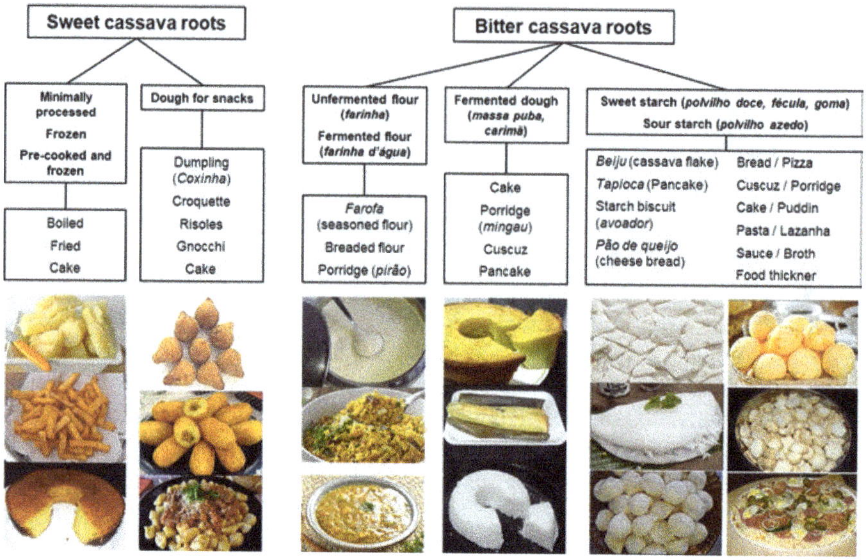

Fig. 7.1 Diversity of products in Brazil made with processed cassava root. (Source: Figure adapted with photos from Alfredo Alves, Joselito Motta, Oliveira et al. (2020a), Podium Alimentos)

on good agronomic practices for growing cassava, including soil preparation, planting, cultivation management, harvesting, and technologies for root postharvest processing.

7.2 Embrapa's Technical Cooperation with Africa: Focus on Cassava

Between 2016 and 2019, *Embrapa Mandioca e Fruticultura*, which also works with tropical fruits (banana, pineapple, papaya, passion fruit, and citrus), received a total of 35 delegations, involving 205 international visitors from 62 institutions and 36 countries from American, African, Asian, and European continents (Fig. 7.2). These visits were aimed at knowledge exchange and technology transfer activities, and most of them (62%) came from the African continent, with greater interest in cassava crop (82%) (Fig. 7.3). For these visits, the trainees explored a variety of topics, including (1) postharvest processing of cassava root and starch, aiming to add value and new products from Brazil that were not yet adopted or adjusted to Africa; (2) good agronomic practices for cassava cultivation, covering technologies from planting to harvest; and (3) production and distribution of cassava seeds with good genetic and phytosanitary quality, focusing on techniques for cleaning planting material via micropropagation, multiplication, and seed distribution. The main

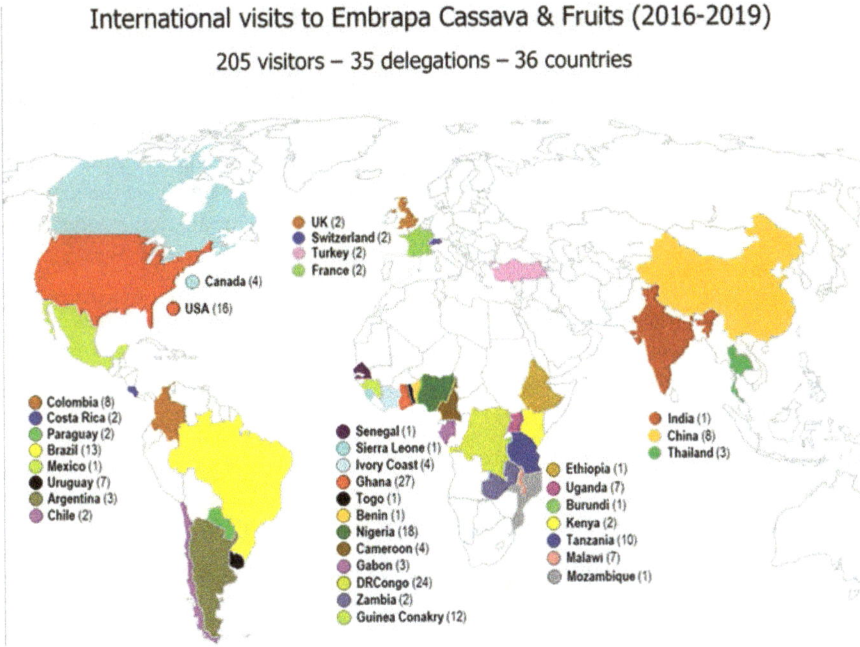

Fig. 7.2 International visits received by *Embrapa Mandioca e Fruticultura* for trainings and technical visits (2016–2019). (Source: NRI-Embrapa Mandioca e Fruticultura (2019))

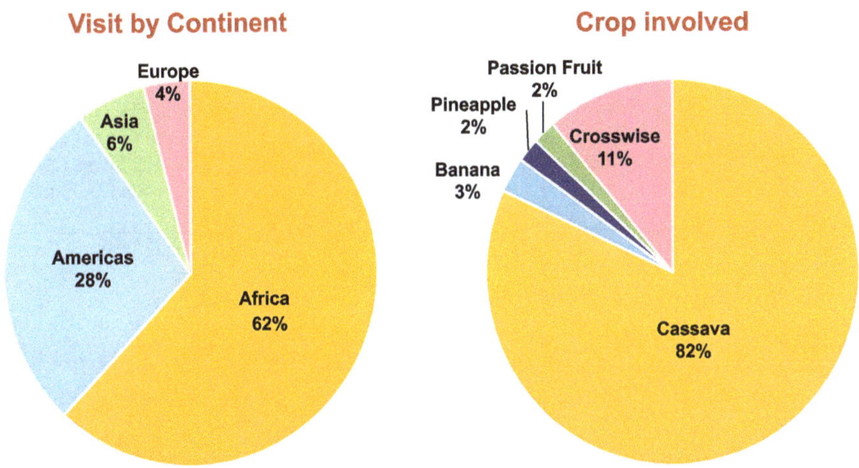

Fig. 7.3 Distribution of international trainings and technical visits by continent and crop of interest. (Source: NRI-Embrapa Mandioca e Fruticultura (2019))

target audience for these trainings were young entrepreneurs (technicians, farmers, and cassava processors) working on their own or in national R&D or private institutions.

7.2.1 *Technical Training for Young Africans*

In September 2017, the Brazil Africa Institute (IBRAF) launched the Youth Technical Training Program (YTTP) aiming to promote knowledge transfer and develop skills of young Africans in various sectors, including agriculture (AfDB 2017). Financed by the African Development Bank, this program chose the cassava production chain on which to focus the initial training on "propagation, production, and processing of cassava" for young Africans. *Embrapa Mandioca e Fruticultura* was invited to plan and execute this training (Table 7.1).

The training program included theoretical classes, followed by practices in the field and laboratories, and visits to cassava processing units and cassava farmers cooperatives located in the State of Bahia (Fig. 7.4). The main topics covered by trainings included:

1. Postharvest processing of cassava root and starch
2. Good practices for cassava cultivation
3. Production and distribution of disease-free cassava seeds

More details about the training's agenda are available in Table 7.2.

Table 7.1 Trainings for young Africans conducted by *Embrapa Mandioca e Fruticultura* and promoted by the Youth Technical Training Program, 2017–2019

Training	Period (# of days)	# of trainees	# of countries	Country (# of people)
Training on cassava propagation, production, and processing for young Africans	23 Oct to 17 Nov 2017 (27 days)	28 17 women 11 men	14	Benin (1), Burundi (1), Cameroon (1), Congo DR (5), Ghana (1), Ivory Coast (3), Malawi (1), Mozambique (1), Nigeria (4), Senegal (1), Sierra Leone (1), Tanzania (4), Uganda (2), Zambia (2)
Workshop on cassava postharvest processing for young Africans	19 to 21 Nov 2018 (3 days)	8 4 women 4 men	4	Cameroon (3), Ghana (3), Ivory Coast (1), Nigeria (1)
Training in cultivation and cassava processing for young Africans	4 to 8 Nov 2019 (5 days)	26 13 women 13 men	3	Ghana (13), Malawi (6), Nigeria (7)
Total	**2017–2019 (35 days)**	**62 34 women 28 men**	**14**	**Benin (1), Burundi (1), Cameroon (4), Congo DR (5), Ghana (17), Ivory Coast (4), Malawi (7), Mozambique (1), Nigeria (12), Senegal (1), Sierra Leone (1), Tanzania (4), Uganda (2), Zambia (2)**

Source: *NRI-Embrapa Mandioca e Fruticultura (2019)*

Fig. 7.4 Training in cassava propagation, production, and processing for young Africans at Embrapa. (**a**) Practical class in the field, visiting a collection of cassava varieties and field instructions for cultivation; (**b**) laboratory practices (micropropagation and multiplication techniques for cleaning planting material); (**c**) training in a cassava processing unit for preparing tapioca from starch extracted from the roots; and (**d**) visit to street market (diversity of cassava products for sale). Cruz das Almas-BA, Brazil, 2017. (Photos: Embrapa Cassava & Fruits)

7.2.2 Increasing Performance of the Cassava Industry in West and Central Africa (IPCI Project)

Funded by the International Fund for Agricultural Development (IFAD), this project ran from 2014 to 2017 aiming to provide technical expertise and support to a range of national cassava projects across West and Central Africa (WCA), including Gabon, Ghana, Nigeria, and the Republic of the Congo. As part of this project, IFAD established a partnership for *Embrapa Mandioca e Fruticultura* to engage in IPCI activities and conducted four primary activities in 2017 (NRI 2017):

1. *Promotion of novel and innovative Brazilian cassava products*. In September 2017, Embrapa welcomed a delegation of ten people, consisting of representatives from IFAD and the project coordinating institutions, and chairpersons from cassava processing units, from Gabon, Ghana, and Nigeria for (1) visits to laboratories and experimental areas of *Embrapa Mandioca e Fruticultura*; (2) a visit to cassava production and processing areas in Bahia State; (3) a visit to Embrapa headquarters to discuss opportunities for international technical cooperation between Embrapa and African countries; and (4) technical meetings on cassava processing with the Embrapa cassava technical team and discussion of IPCI project. On this technical visit to Brazil, the IFAD partner researchers explored,

Table 7.2 Content of the training on "Cassava Propagation, Production, and Processing for Young Africans"

Theoretical classes	Field and laboratory practices at Embrapa	Visits to processing/market units and cassava farmers cooperatives
Embrapa's Cassava Breeding Program Cassava agronomic aspects Climate change and cassava water use efficiency Cassava pests and diseases Cassava weed management Reniva project (multiplication and distribution of disease-free cassava seeds) Toxicity and processing cassava Cultivation technologies for small cassava producers	Visit to Embrapa's in vitro and field cassava genebanks Field practices on soil management Visit to physiology and climatology laboratories Visits to entomology, phytopathology, and virology laboratories Root processing practices: farinha production and starch extraction Field practices on cassava cultivation techniques Minimal processing of cassava roots – food technology laboratory Processed product preparation practices	Visits to starch industry in Laje-BA (Bahiamido) and the cassava farmers cooperative in Tancredo Neves-BA (Coopatam) Visits to street markets in Cruz das Almas and Vitória da Conquista-BA (cassava products for sale) Visit to small- and medium-scale *beijus* production units Visits to cassava processing units and Coopasub (Cooperative of Small Farmers in Southwest Bahia) in Vitória da Conquista-BA Visit to Bahia Biofactory (IBB) in Ilhéus-BA: practices on cassava seed production system Practices on culture medium preparation and plant micropropagation at IBB Cassava seed field production: Reniva Project, in Tapera (Cruz das Almas region)

Source: *NRI-Embrapa Mandioca e Fruticultura (2017)*

with Embrapa researchers, Brazilian technologies related to the use and processing of cassava, at different technological levels, from artisanal (small farming) to industrial processing at medium and large scales. The visits occurred in two cassava-producing regions with different profiles regarding the use and processing of cassava roots: (1) the northeast region, where small farmers predominate and the roots are processed mainly for production of flour and starch derivatives; and (2) the southern region, where cassava is grown on a large scale for production and transformation of starch.

2. *Embrapa publications translated.* Embrapa maintains a large stock of publications and gray literature on cassava. With respect to IPCI, literature relating to postharvest and processing aspects of cassava might be especially useful. However, most of these publications are written in Portuguese. Translating these publications to English and French could have a significant impact on the African cassava research environment. Embrapa identified the most appropriate publications for the areas of utilization and postharvest processing of cassava and provided for the translation of 11 publications into French and English. In addition, Embrapa selected multimedia material (videos and documentaries) reedited with English and French subtitles.

3. *Survey of equipment and machinery for cassava processing*. Brazilian manufacturers have developed a wide range of cassava processing machines which might be useful to the cassava processing sector in Africa. Usually, contacts between cassava processors in the target countries and suppliers of equipment are ad hoc. Embrapa made a brief survey of Brazilian suppliers of cassava processing equipment and provided a list of equipment, specifications, and prices, with the key elements translated into English and French (more information available in Table 7.3).
4. *Technical visit of Embrapa experts to target countries*. Three Embrapa cassava experts visited Nigeria and Ghana, where they learned about IFAD's ongoing activities and its projects related to cassava processing through visits to IPCI partner institutions. On the part of farmers and processors, the mission visited cassava production fields, gari processing centers, and family cassava processing units, where visitors gave practical demonstrations on how to prepare *tapioca* from cassava starch (Fig. 7.5a, b). These visits were important to understand the few alternatives for processing cassava roots in those countries, where they concentrate, predominantly, on gari production. The great interest shown by local farmers and processors in the preparation of *tapioca*, for example, highlights the technological gap and potential growth areas for strengthening technology transfer of processing tools from Brazil to Africa.

The trip culminated with "CassavaTech" in Lagos, Nigeria, where the Embrapa team organized an exhibition of cassava products produced in Brazil with practical demonstrations on how to prepare *tapioca* (Fig. 7.5c) and a special session on "Brazilian Technologies for Cassava: What Africa can learn from the Brazilian Cassava Industry."

7.3 Major Innovations for Cassava Root Processing

Cassava roots are processed in various ways to prevent postharvest deterioration, reduce root toxicity, improve the palatability of derived products, and increase the commercial value of cassava.

Based on the training evaluations and the interest shown by the participants, we selected some technological innovations already in use in Brazil and ready for adaptation for Africa, with great potential for adoption and market opening for new products.

The innovations refer to products obtained from the processing of sweet cassava roots and from bitter cassava roots and starch. It is important to note that all products/technologies applied to bitter cassava can also be used in sweet cassava, but the reverse is not always true.

Table 7.3 Information about Brazilian manufacturers and their equipment for processing cassava roots

Manufacturer (website)	Equipment	Web link showing machines operation/catalog
Fankorte – Indústria de Máquinas (www.fankorte.com.br)	Root washer (skin removal) Root peeler Root cutter and sticks maker	https://www.youtube.com/watch?v=NVe9gs5NsPU https://www.youtube.com/watch?v=7zP_2QYWNK4
Jair Pomim (www.youtube.com/channel/UCpEClMBFNwAZoX7i4hnnRHQ)	Mini machine for skin removal and peeling roots Slicing, shredding, and dough maker machine	https://www.youtube.com/watch?v=T23FxjU7v-I https://www.youtube.com/watch?v=rcS70MVGG5g
Maringá Torneadora (https://www.youtube.com/channel/UCrQZ9gbaTEcbXr8ZMIM6EFw)	Mechanized oven for toasted cassava flour	https://www.youtube.com/watch?v=rWfYOnyFdSY
Chapadão Máquinas (www.chapadaomaquinas.com)	All cassava flour processing machines	https://www.youtube.com/watch?v=odXYu900b3w
VMAQ Indústria de Máquinas Especiais (https://vmaq.ind.br/maquinas/)	All the equipment to build a factory for industrial production of starch (polvilho) and cassava flour	https://www.youtube.com/user/vmaqindustria

Fig. 7.5 Practical demonstrations of how to prepare "tapioca" from cassava starch in the District of Baara, Abeokuta (**a** and **b**), and in Lagos, Nigeria, at a CassavaTech event (**c**) in 2017. (Photo credits: Alfredo Alves and Ben Bennet)

7.3.1 Sweet Cassava Processing

For sweet cassava, we highlight cassava minimally processed, frozen cassava, pre-cooked and frozen cassava, cassava chips, and cassava dough for snacks. Oliveira et al. (2020a) describe in detail the preparation steps of these products, which must be carried out following good manufacturing practices.

7.3.1.1 Cassava Minimally Processed

Minimally processed cassava is an alternative to extend the root supply period, providing a product that is easy to prepare and more convenient for consumption (Viana et al. 2010). Minimal processing needs to be carried out in a cold environment (10–15 °C). If this is not possible, use water between 5 °C and 10 °C for root washing and sanitization. Basically, it consists of root reception, selection, washing, sanitizing, cutting, peeling, sanitizing, vacuum sealed packaging, and storage at 5 °C (Fig. 7.6).

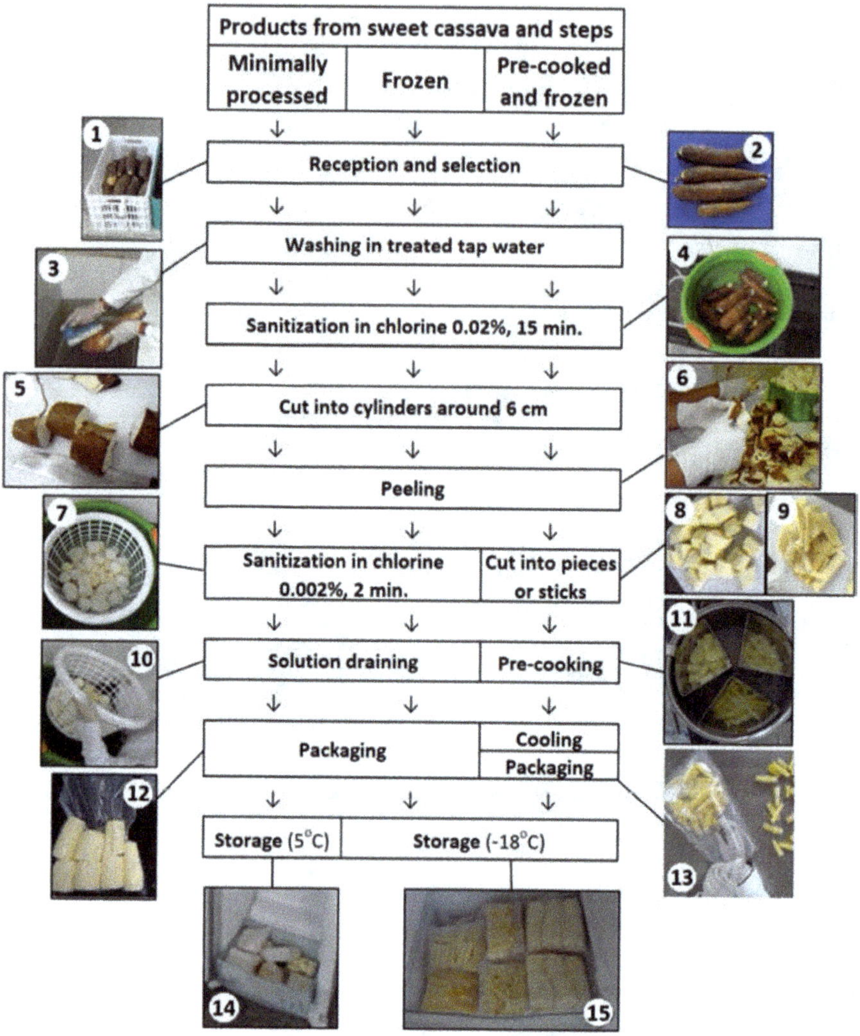

Fig. 7.6 Steps for preparing minimally processed cassava, frozen, and precooked and frozen. (Source: Figure adapted with photos from Oliveira et al. (2020a), Oliveira et al. (2020b), and Viana et al. (2010))

Some steps to prepare minimally processed cassava are also present in the protocol of all products, such as reception and selection, washing, sanitization, peeling, and packaging of the roots, where some good practices must be followed.

Reception and washing of fresh roots To minimize cross contamination, the reception and washing areas for fresh roots must be separated from the internal processing areas. The roots must be harvested on the same day of processing or the day before processing and stored at night. Usually, fresh sweet cassava roots are received in boxes or bags (Fig. 7.6(1)) and washed in treated tap water with a brush to remove soil and dirt residues (Fig. 7.6(3)).

Sanitization Sanitization is performed after washing the roots (with and without peel) and on the materials/tools used in the root processing. This step aims to reduce the number of deteriorating microorganisms and eliminate pathogens (which cause diseases). For each material to be sanitized, different chlorine solutions are used (Table 7.4).

Peeling the roots Peeling the roots must be careful as it can damage the tissues and accelerate the browning process, called physiological and/or microbial deterioration, and increase the water loss. Peeling sweet cassava roots for culinary use is usually done manually with a stainless-steel knife or with suitable peeling equipment. After removing the root tips, if the peeling is carried out manually, the cylinders are cut to approximately 6 cm in length and then peeled. If peeling is done with equipment, cassava is cut into cylinders after peeling. All the peel must be removed to improve the appearance of the product and avoid darkening (Fig. 7.7).

Table 7.4 Sanitization steps performed in the postharvest processing of cassava

Sanitization of	Procedure	How to prepare 1 L of sanitizing solution
Roots with bark	Immersion in *0.02% chlorine solution* for 15 minutes	*0.02% chlorine*: 10 mL of 2% sodium hypochlorite (commercial) in 1 L of water = 200 mg of chlorine/L *0.002% chlorine*: 1 mL of 2% sodium hypochlorite (commercial) in 1 L of water = 20 mg of chlorine/L *0.01% chlorine*: 5 mL of 2% sodium hypochlorite (commercial) in 1 L of water = 100 mg of chlorine/L
Peeled roots	Immersion in *0.002% chlorine solution* for 2 minutes	
Materials used in the processing steps (knives, table/bench surface, boards, sieves, etc.)	Immersion in *0.01% chlorine solution* for 2 minutes or spray with *0.02% chlorine solution*	

Source: Adapted from Oliveira et al. 2020a

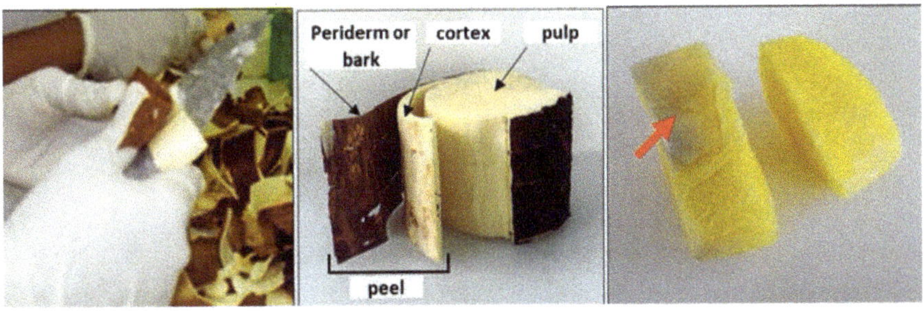

Fig. 7.7 Peeling sweet cassava roots. All peel (periderm and cortex) must be removed. The presence of peel in precooked root impairs the appearance of the product (red arrow on the right). (Photo credits: Eliseth Viana and Luise Sena)

Packaging the roots Packaging protects the product during transportation, distribution, storage, and handling against damage, shock, vibration, and compression that occur during the entire journey. The low-density polyethylene plastic bag is the most commonly used packaging for processed cassava products, as it is resistant to impact and a good water barrier. However, some products, such as flour, may use kraft paper packaging, molded as a box or bag; this material has the advantage of being recyclable and biodegradable. For fried products, such as chips, polypropylene and aluminum plastic film are recommended to prevent oil oxidation and maintain the crispness and flavor of the chips for a longer time. However, it is common to use polypropylene plastic bags for the packaging of fried chips, as they are less expensive, sold in small quantities, and easier to find. The polypropylene packaging is transparent and gives greater visibility to the product, but the chips will have a shorter shelf life due to oil oxidation.

7.3.1.2 Frozen Cassava

Frozen cassava is sold in whole cylinders from 5 to 7 cm in length, in pieces, or crushed to make cakes. The processing steps are the same as for minimally processed cassava with the exception of storage that is carried out at −18 °C in a freezer or cold room (Fig. 7.6).

To obtain frozen cassava, the roots must first be selected, excluding those with hard (fibrous) parts, spots, rot, or other problems (Fig. 7.8). The first selection of the roots usually occurs in the field at harvest time. To evaluate the root quality, a sample of the harvested roots that represents the variety to be processed is cooked. Only the group roots that present desirable quality (e.g., cooking in 30 minutes or less) should be processed. Roots that cook in more time must not be processed, even under pressure. Processing does not improve the quality of the processed root, so it is essential to process good quality roots to obtain a good product (Oliveira et al. 2020b).

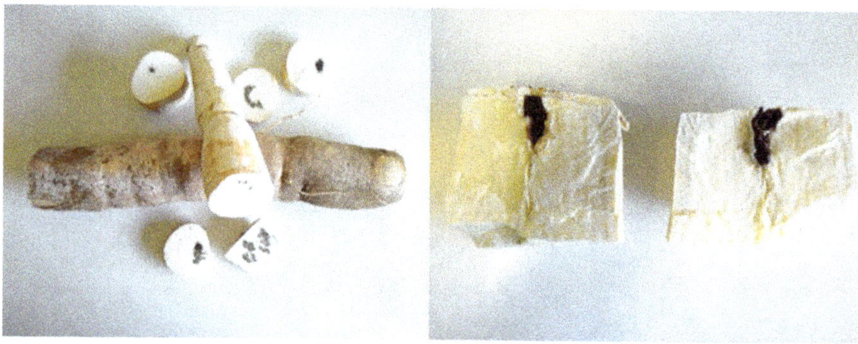

Fig. 7.8 Cassava roots unsuitable for processing due to the presence of dark spots. (Photo credits: Luciana Alves de Oliveira)

After the second sanitization, the sanitizing solution is drained, and the cylinders are packed in more resistant polyethylene packages. Frozen cassava can be sold in portions from 200 g to 2 kg, depending on the consumer market. The packages are sealed and placed at −18°C in a freezer. Vacuum packaging is not recommended as frozen cassava can pierce the packaging.

If the processing is carried out properly, the product will last for at least 4–6 months. The roots must be placed for cooking while still frozen, and this information must appear clearly on the packaging.

Losses in the processing of frozen cassava, such as peel and tips, vary from 25% to 30% of the total weight of the roots. The tips and small pieces of cassava can be used for the preparation of other products, such as dough for snacks.

7.3.1.3 Precooked and Frozen Cassava

Precooked and frozen cassava is sold in the form of sticks or pieces. In this processing, the initial steps are the same as for frozen cassava (root reception, selection, washing in water, sanitation, cut in cylinders, peeling), followed by cutting into sticks or pieces, precooking, cooling, packaging, and storage at −18°C in a freezer or cold room (Fig. 7.6).

The stick is always cut longitudinally to the length of the root in the direction of the fibers. If the stick is cut across the fibers, the percentage of breakage is greater and the product length will be shorter (Fig. 7.9). The stick is cut 1 cm on the side and 5–6 cm long. For cutting the roots into pieces, the cylinder can be cut into 4–8

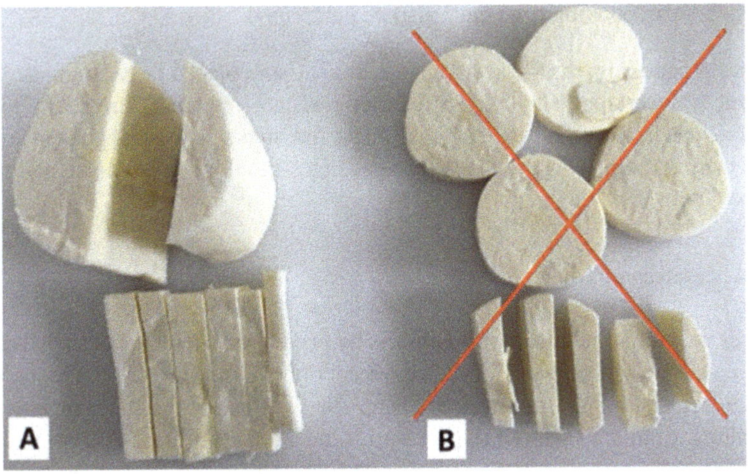

Fig. 7.9 Photo (**a**) shows the proper way to cut cassava lengthwise and with regard to thickness. Photo (**b**) shows improper cutting of cassava for sale, which will diminish its safety and shelf life. (Photo credit: Luise de Oliveira Sena)

Table 7.5 Tests performed to determine the appropriate precooking time for two cassava varieties processed as sticks

Precooking time	After frying	
	Variety 1	Variety 2
No precooking	Taste of raw cassava	Taste of raw cassava
2 minutes	Taste of raw cassava	Adequate flavor and texture
4 minutes	Taste of raw cassava	Oil soaked
6 minutes	Adequate flavor and texture	Extremely oil soaked

Source: Oliveira et al. (2020a)

parts. Imperfect sticks, tips, and leftover roots can be used to produce dough for snacks.

After cutting, the precooking starts with 2 L of water for each kilogram of cassava root and 2% salt added to the water (20 g of salt per 1 kg of cassava root) to accentuate the flavor. The cooking water must be changed after three uses. The sticks or pieces of cassava are placed in the water once it is boiling.

The precooking time should be tested for each variety used. For example, precooking can be tested in three time periods: 2, 4, and 6 minutes. For each test, use 250 g of cassava sticks and 0.5 L of water with 5 g of salt. The precooked sticks should be frozen when submerged in water, boiled for the required time, and then fried to verify the quality of the final product. Table 7.5 shows the results of the test performed with two cassava varieties. Each variety was harvested and tested on the same day and at the same age.

Precooking time can vary for the same variety, depending on the root age and environmental conditions (fertilization, rainfall, plant nutrition, soil, drought). The precooking time for a piece of root is longer than that for a root stick. Thus, tests to confirm best precooking time must also be performed.

After precooking, the sticks should be drained and placed in packs made with a resistant (thicker) polyethylene or another resistant material. The sticks tend to adhere to each other. To prevent this, place the sticks in a disorganized manner inside the packaging (Fig. 7.6(13)). The packages should then be sealed and placed in a freezer at −18 °C.

7.3.1.4 Cassava Chips

Another way of adding value to cassava is by producing fried snacks such as chips. The preparation of the chips includes the reception and selection of the roots, washing in water, sanitizing, peeling, sanitizing, slicing, soaking the slices in water, frying, salting, draining surface oil, packaging, and storage (Fig. 7.10).

Peeling the entire root is important to reduce leftover tips. After peeling, the peeled roots should be submerged in treated water to remove any adhering skin. The water used in this stage can be reused for initial root washing.

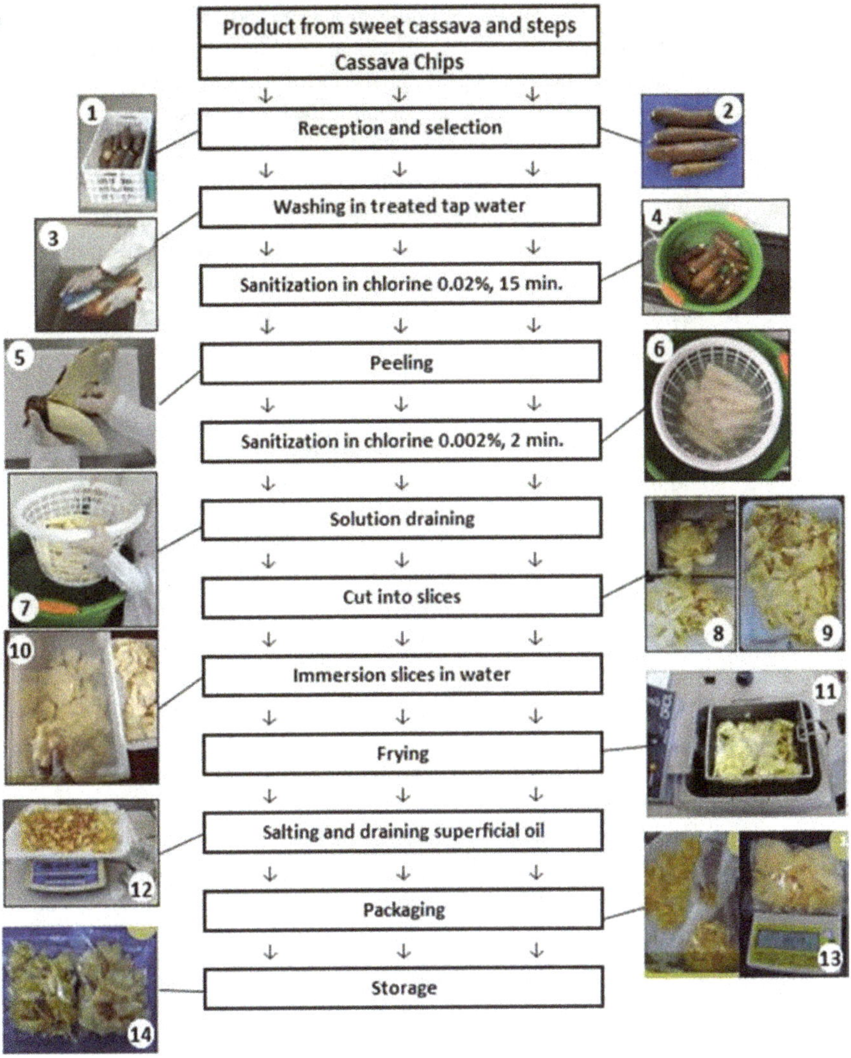

Fig. 7.10 Steps for preparing cassava chips. (Source: Figure adapted with photos from Oliveira et al. (2020a, b)

The root is cut into slices approximately 0.8 mm thick. The thickness of the slice is especially important for the quality of the product (crispness). If the slice is too thick, the chips may become hard. There is cassava slicing equipment available on the market for this purpose. Slicing can occur directly into the fryer; if not available, slices must remain immersed in water to prevent them from sticking. After soaking, the water must be drained and the slices fried. Very thick cassava roots are not suitable for chips as they crack during peeling or slicing (Oliveira and Godoy 2011).

The frying oil temperature should be between 170 °C and 180 °C and not exceed 190 °C to avoid impairing the oil quality. The quality of the oil used for frying will influence the quality and shelf life of the finished product. Oil degradation will be greater the longer the period of its use. To monitor the quality and disposal point of the frying oil, there are quick tests and evaluation equipment on the market. Palm oil has higher oxidative stability (less degradation) than cotton, sunflower, and soybean oils; therefore, it is the most suitable for the production of fried products.

After frying, excess oil is drained and the chips are spread over a sheet of paper to again drain the excess. The product is salted with 1% salt: 10 g for every 1 kg of product. The addition of spices (oregano, chili powder, parsley, onion, among others) can be added at this time also. If the additional seasoning contains salt, the amount of added salt must be decreased.

The slices should be quickly packaged to prevent moisture absorption. The product should be packed in polypropylene bags, in portions of 40–80 g, stored in a dark place, and should be consumed within 15 days.

Chips may vary in quality due to factors such as the quality of the raw material and oil used, inequalities in chip thickness, temperature, and frying time. The color of the final product depends on the color of the cassava variety used, which may be white, cream, yellow, or pink.

7.3.1.5 Cassava Dough for Snacks

For the preparation of dough for snacks, the first steps are the same as for frozen cassava, followed by the steps of cutting the root into four pieces, central fiber removal, cooking, freezing, grinding, salting, shaping, packaging, freezing, and storage at −18 °C in a freezer or cold room (Fig. 7.11). If processed properly, the product will last at least 6 months. Before starting the dough preparation, one batch of roots should be used to evaluate the cooking time of the variety used.

The cooking water is drained (Fig. 7.11(10)), and after the cooked roots return to room temperature, they should be cooled at a temperature between 4 °C (refrigerator) and −18 °C (freezer) (Fig. 7.10(11)) for 24 hours. The cooked root, still frozen, is ground in a meat grinder (Fig. 7.11(12)). Then the crushed dough is mixed with salt, using a ratio of 10 g salt to 1 kg dough (1% salt) in a mixer (Fig. 7.11(13)), adding the salt gradually while mixing the dough.

The dough can be placed in a sausage maker (Fig. 7.11(14)) to make croquettes. The machine's screw system should rotate slowly so that the texture is uniform, and the dough does not acquire air bubbles. When the dough begins to exit the plastic section of the equipment, use hands to press the plastic section outlet (Fig. 7.11(15), red arrow) while the screw system turns. It is important to exert pressure during this step so that all air is removed from the dough. The tray that will receive the dough must be greased with a little oil so the dough does not stick to the tray.

The croquettes must quickly be cut to the same size using a knife (Fig. 7.11(15)) so that the dough does not break. The knife should be rinsed in water to facilitate

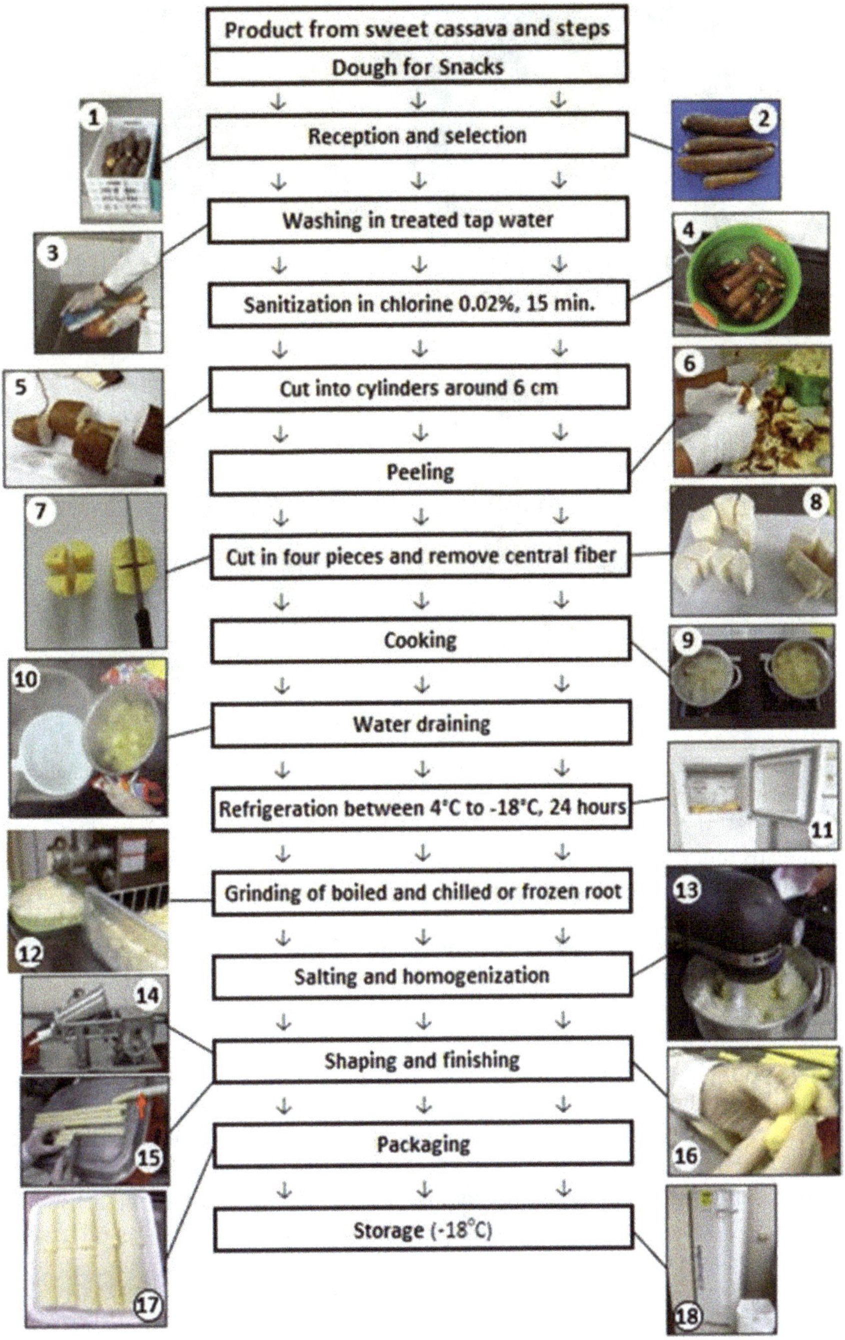

Fig. 7.11 Steps for preparing cassava dough for snacks. (Source: Figure adapted with photos from Oliveira et al. (2020a))

Fig. 7.12 Frozen (**a**), frying (**b**), and fried (**c**) croquettes and "coxinhas" made from stuffed sweet cassava dough. (Photo credits: Luise de Oliveira Sena)

cutting. To finalize the savory pastry, dip fingertips in water and mold the cut ends with moistened fingers (Fig. 7.11(16)) to round the ends of the croquettes.

Croquettes are sold frozen or pre-fried (180 °C for 35 seconds) and frozen. Fry the frozen croquettes in oil at 170 °C for 3–4 minutes in a covered pan, as the croquettes may explode if the dough is too wet.

In addition to salt, other spices can be added to the dough, such as onion, cilantro, basil, oregano, cheese, and parsley. The dough can also be used to make savory foods in different formats ("coxinha," dumpling, rissole, gnocchi) and with desired fillings (bacon, shrimp, ground beef, chicken, cheese, among others) (Fig. 7.12). Savory pastries tend to be quite sticky and need to be breaded. The flour used for breading also removes the sticky quality of the pastries and becomes an oil absorption barrier during frying, which increases the crispness of the product.

7.3.2 Bitter Cassava Processing

In Brazil, bitter cassava roots are processed to produce **flour** (*farinha de mandioca*) and **starch** (known in Brazil as *polvilho*, *fécula*, or *goma*), which can be further processed into a series of products. Below, we will highlight ***beiju*** **(cassava flake)**, ***tapioca*** **(stuffed beiju)**, ***avoador*** **(starch biscuit)**, and **cheese bread**.

7.3.2.1 Flour Processing

Dry flour is consumed throughout Brazil, and its production involves the following steps: washing and peeling the roots, grating, pressing, crumbling, sieving, roasting, sifting, packaging, and storage (Fig. 7.13). Fermented flour, known in Brazil as *farinha d'água* or *farinha puba*, is widely consumed in the Amazon region (North Brazil). Its production involves the same steps as dry flour, except that, before being grated, the roots are macerated (softening of the roots in a fermentative process in water, known as *pubagem*) for 4 days. After *pubagem*, the same steps as dry flour should be followed (Fig. 7.13).

To prepare the flour, it is important to use the same variety and the same age of cassava roots. Root washing prior to manual peeling (or maceration for fermented flour) is recommended (Fig. 7.13(1)). Processing should begin immediately after harvest. Roots must be completely peeled where the bark is not white. Partial peeling can influence the quality of the flour and lead to dark spots and/or browning in the final product.

The peeled root should not be placed directly on the ground but rather in plastic boxes and washed with tap treated water to remove sticky peels, sand, and other adhered dirt.

The next step is root grating, which is done using an electric grater, also called *cevador* or *caititu* (Fig. 7.13(6)). The grater has a cylinder, preferably with serrated stainless-steel blades (Fig. 7.13(4, 5), red arrow) that can be replaced as needed.

Excess water in the grated dough must be removed before roasting using manual (Fig. 7.13(9)) or hydraulic pressing equipment. The grated root dough is placed in raffia bags (Fig. 7.13(7)) or on fine mesh nylon screens and pressed onto pallets (Fig. 7.13(8)) to remove excess water (also called *manipueira*. The pressing time depends on the amount of crushed mass and the equipment used. *Manipueira* and cassava peel can be used in animal feed, fertilization, and irrigation.

The dough becomes compact due to compression and needs to be crumbled. This procedure is performed using the same root grater (Fig. 7.13(6)).

To get a fine dry flour (which is preferred in northeast Brazil), it is recommended to pass the crumbed dough through a hand or mechanical sieve (Fig. 7.13(10)) with a galvanized 26–28 mesh to standardize the grain size of the flour and remove the fibers (central root fiber) before drying and toasting. The higher the mesh number, the finer the flour obtained.

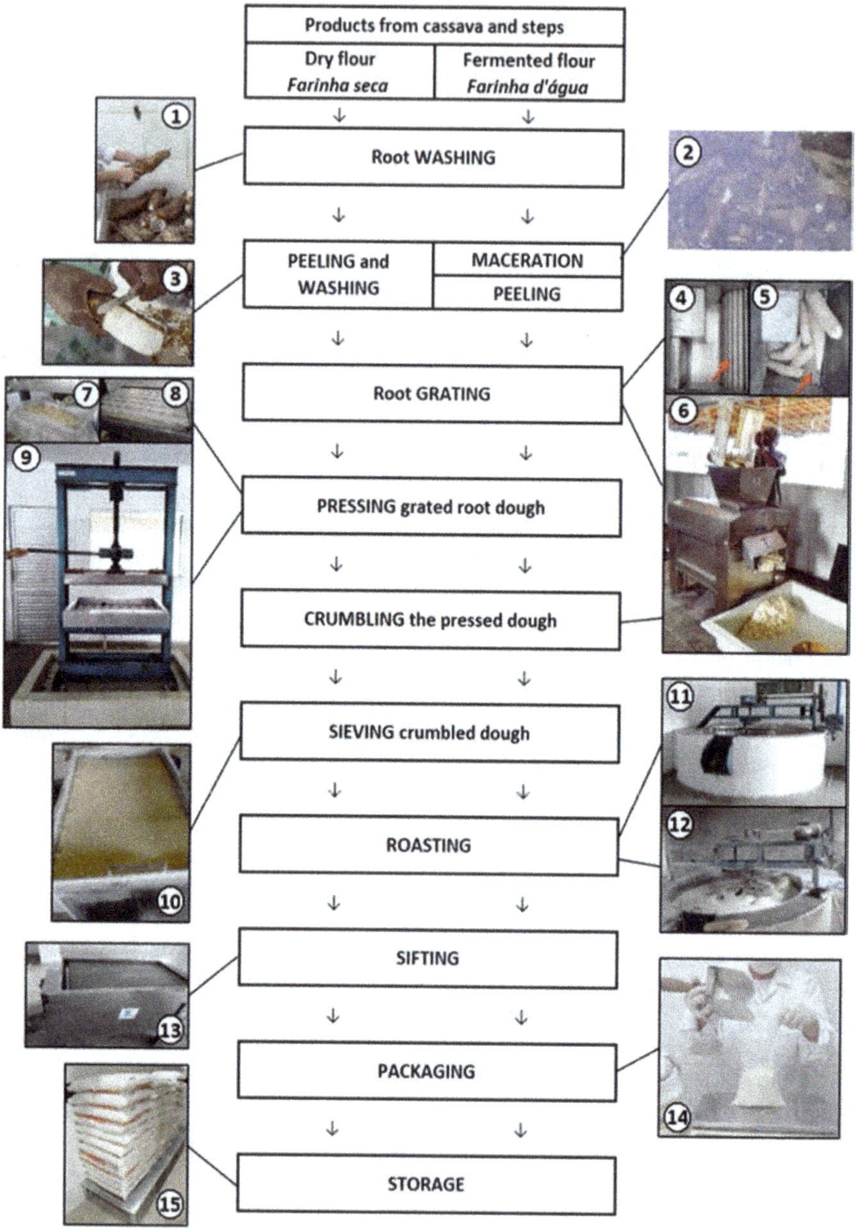

Fig. 7.13 Steps for preparing dry and fermented cassava flours. (Source: Figure adapted with photos from Oliveira et al. (2020a) and SENAR (2018))

The dough is manually placed in the oven using a scraper (Fig. 7.13(11)). The dough should be added slowly into the oven with the stalks moving so as not to tangle.

Initially, the dough should be roasted for 20–30 minutes at 90 °C. After that time, increase the oven temperature to approximately 160 °C for up to 3 hours to reduce moisture (Fig. 7.13(12)). After roasting, the flour must be cooled and sifted to evaporate the remaining moisture (Fig. 7.13(13)). The flour grain size depends on the demand of a particular consumer market.

To keep the flour crispy, packaging for storage and sale is important. When flour is ready and cold, 50 kg plastic bags are recommended. For smaller portions, 1 kg packages of polypropylene are preferred (Fig. 7.13(14)), which preserves the low moisture of the flour. Polyethylene keeps the product crisp for a shorter time.

The flour must be stored in a dry and ventilated place with screened windows. It is recommended that the bags be stacked on pallets or crates, leaving room for air circulation between the packs (Fig. 7.13(15)). According to Brazilian legislation, cassava flour must have a maximum humidity of 13%.

*Gari **vs** farinha d'água* The most consumed processed cassava product in Brazil is unfermented cassava flour, which is not made in other regions of the world. Outside Brazil, the main cassava product is **gari**. *Gari* was introduced in Africa in the early nineteenth century after freed slaves returned to the African continent. Today, it is the main cassava product in sub-Saharan Africa, representing 70% of the cassava consumed in Nigeria. Meanwhile, in Brazil, *farinha d'água* is produced, almost exclusively, in the Amazon region (Folegatti et al. 2005; Bechoff et al. 2019; Ndjouenkeu et al. 2021).

Both flours are made from fermented cassava. However, the fermentation of *farinha d'água* occurs in anaerobic conditions by immersing the whole root in water, while gari fermentation is aerobic: the grated root mass is pressed slowly for 2–6 days (Fig. 7.14). Gari also includes a step for "garification," which combines stages of gelatinization (starch cooking) and gari roasting, which affect the physicochemical and rheological properties of the starch, providing this food its unique sensory and functional characteristics (Folegatti et al. 2005).

Massa puba **(puba dough) or** *carimã* *Massa puba* is produced from the spontaneous fermentation of fresh cassava roots (with or without peel). The roots are placed in water for approximately 5 days until they soften and begin to release the bark. Then, they are crushed in sieves and washed until only the fibers remain. The separated mass must be washed several times and dried in the sun or in dryers until a humidity of 50% is reached for wet puba (or *puba úmida*) and 13% for dry puba (or *puba seca*) (Ferreira Filho et al. 2013). *Carimã* is a fine flour made by crushing dried fermented roots. *Puba* and *carimã* are used in the preparation of cakes, porridge, couscous, and other dishes.

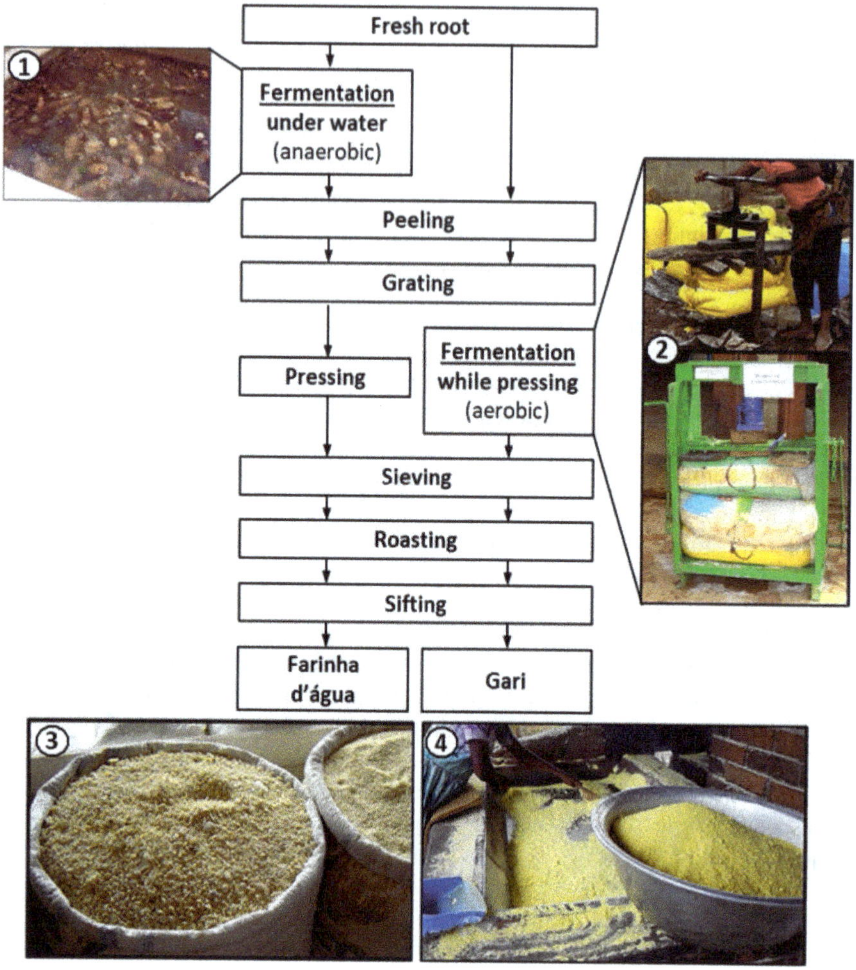

Fig. 7.14 The steps to prepare *farinha d'água* and *gari*, including the processing differences between the two products. (1) With *farinha d'água*, fermentation is anaerobic and the whole roots are immersed in water, (2) while in gari, the fermentation is aerobic and the grated root mass is left to ferment during pressing. (Photos: Alfredo Alves (2-4) and Folegatti et al. 2005 (1))

7.3.2.2 Small-Scale Starch Extraction ("Polvilho Doce" or "fécula")

In the past, the cassava starch was considered the by-product of flour production. Due to the increased demand by the consumer markets for cassava products, the starch has been used to manufacture starch biscuits, tapioca, and cheese bread (SENAR 2018).

The steps for starch extraction (Fig. 7.15) are the same as for dry flour until the grating step. The crushed root dough (Fig. 7.15(6)) should pass through the starch

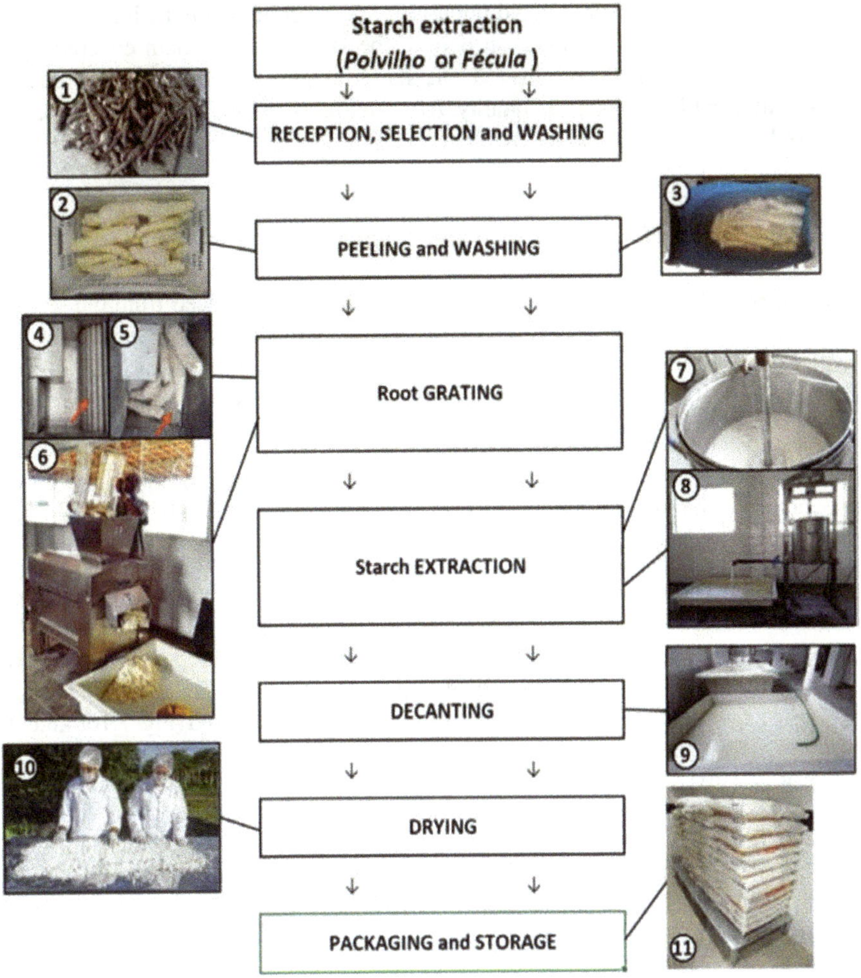

Fig. 7.15 Steps for cassava starch extraction. (Source: Figure adapted with photos from Oliveira et al. (2020a) and SENAR (2018))

extractor (Figs. 15.7 and 15.8) and mix with water while in the starch extractor. As the water drains out, it must be received in a glass fiber trough to separate the starch via decantation (Fig. 7.15(9)). The decanted starch is crumbled and placed to dry in the sun until it reaches a humidity of approximately 13%. The dried starch is then packaged and stored in a cool and airy place. Starch can be stored safely up to 12 months.

Sour starch ("polvilho azedo") The production of sour starch follows the same steps as natural starch, except that in the decantation stage, the starch is left to ferment in the tanks for a period ranging from 15 to 40 days, depending on the ambient temperature, until the product reaches an acidity of approximately 5%. The settling tanks must be under cover to avoid impurities and must be covered with ceramic or fiber tiles. During the fermentation period, the starch must remain covered by a 10 cm layer of water. If not covered, the starch exposed to air acquires a blue or purple color, which reduces its quality. After the fermentation period, the starch is removed from the tanks and taken to dry.

7.3.2.3 "Beijus" (Cassava Flakes) and "Tapiocas" (Stuffed Beiju)

The *beijus*, or tapiocas as they are known in Brazil, are produced from the starch, gum, *fécula*, or *polvilho*. Two types of this product are common: (1) *beijus* roasted on hot plate, crispy, without fillings and (2) soft and folded with various fillings, often called stuffed tapioca, *beiju de tapioca*, or simply tapioca (Motta, 2012).

These products are made with cassava-hydrated starch (*farinha de tapioca* or *polvilho doce úmido*), which is obtained by mixing a part of water with two parts, by weight, of the natural starch (*polvilho doce seco*) and pinches of salt. The mixture is homogenized and sieved to make it thinner.

To make traditional roasted beiju, hydrated starch is spread on a hot plate or heated pan (Fig. 7.16(1)). The drying point of the dough can be identified when the edges come loose and become brittle, like pancakes or a dry crepe. The *beijus* are cut on the hot plate (Fig. 7.15(2)), cooled to room temperature (Fig. 7.16(3)), packed in plastic bags or pots (Fig. 7.16(4)), and sealed, labeled, and stored (Fig. 7.16(4,5,6)).

Colored "Beijus." Traditionally white, *beijus* have been produced in new colors, smells, flavors, and nutrients when the water is replaced by fruit pulps or vegetable extracts. Tasting evaluations found that consumers preferred these products flavored with beets, onion, pineapple, guava, and passion fruit (Fig. 7.17a) (Motta, 2012). To boost cassava sales in schools, producers have also manufactured gum biscuits, cookies, and other products (Fig. 7.17b).

Stuffed tapioca ("beiju de tapioca" or "tapioca recheada") To make tapioca, start by hydrating the natural starch (*polvilho doce* or *tapioca seca*) and adding a part of water to two parts of starch (Fig. 7.18(1)). Mix well with hands until it crumbles and add a pinch of salt. Sieve the hydrated starch (Fig. 7.18(2)) and spread it over a preheated frying pan (Fig. 7.18(3)). To spread the sieved starch, you can use the sieve or your hands. Let the starch heat up until you notice that the grains have stuck to each other, and the starch pancake becomes loose from the pan (around 1 minute) (Fig. 7.18(4)), and turn it upside down, like a pancake. Add any kind of filling (Fig. 7.18(5)) and fold the tapioca in half (Fig. 7.18(6)). Heat the tapioca a bit more on both sides to complete the process. The frying pan does not need to be greased; it should be clean and without grooves. The type of filling depends on consumer preference, such as butter and cream cheese, condensed milk, ham and cheese, meats, and a range of food mixes (Fig. 7.19).

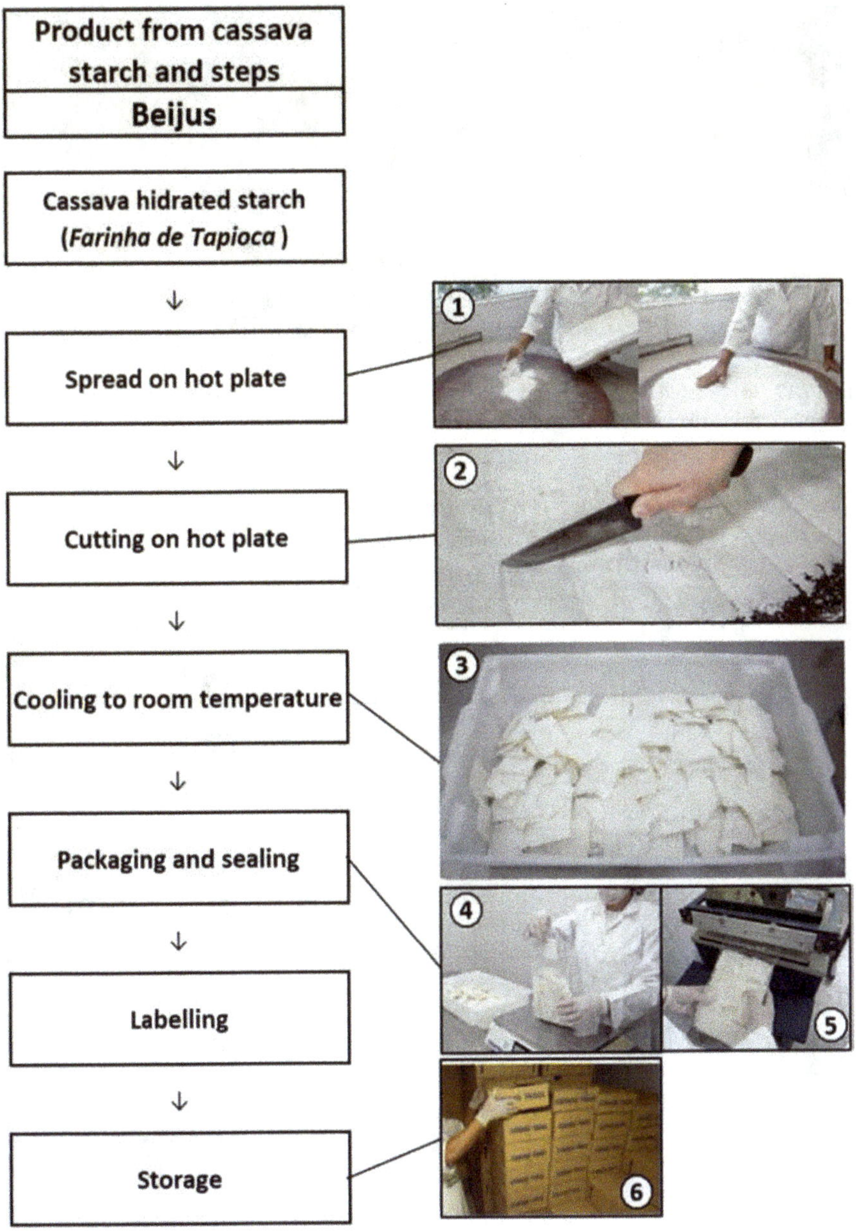

Fig. 7.16 Steps for preparing beijus. (Source: Figure adapted with photos from SENAR (2018))

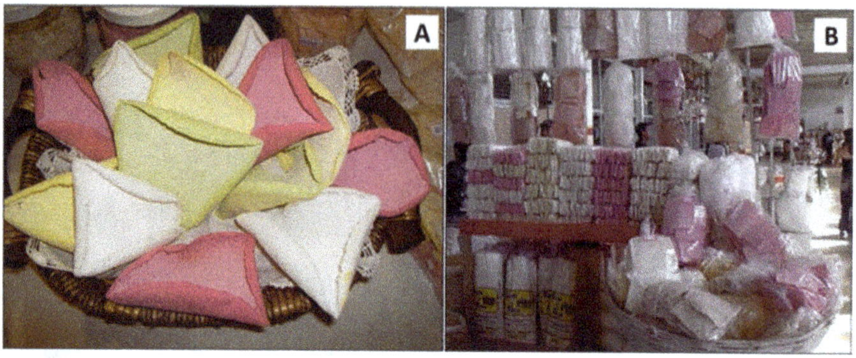

Fig. 7.17 (**a**) Colored *beijus* prepared with fruits and vegetables. (**b**) Diversity of *beijus* for sale at the municipal market in Cruz das Almas, Brazil. (Photo credits: Joselito Motta)

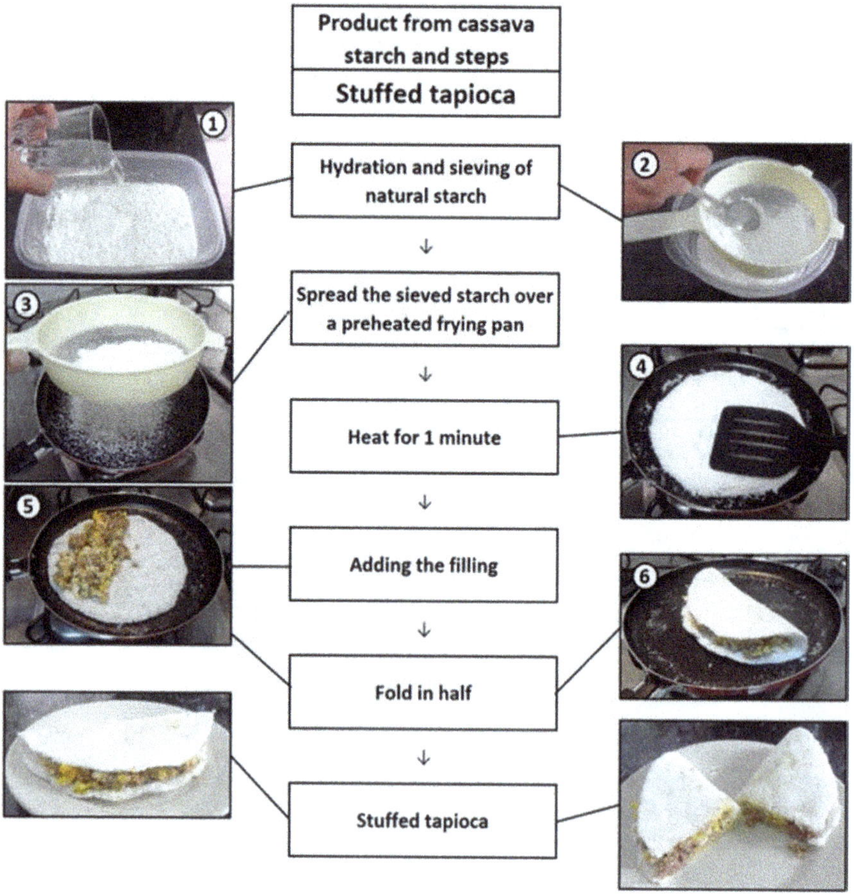

Fig. 7.18 Steps for stuffed tapioca. (Photo credits: Alfredo Alves)

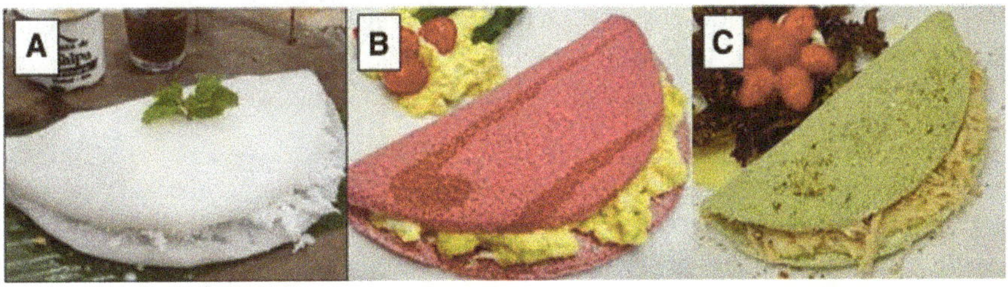

Fig. 7.19 Tapioca with different types of filling: (**a**) plain tapioca with grated coconut filling; (**b**) beet tapioca with scrambled egg filling; (**c**) cabbage tapioca with shredded chicken and mozzarella cheese filling. (Photo credits: Joselito Motta)

Fig. 7.20 Diversity of avoador biscuits and other cassava starch snacks. (Photo credit: Joselito Motta)

7.3.2.4 Avoador (Starch Biscuit)

Avoador, often referred to by its generic name *biscoito de polvilho*, but also known as *peta* or *biscoito avoador*, is a type of snack food in Brazilian cuisine. It is found mainly in the states of Minas Gerais, Bahia, and Rio de Janeiro, but is also consumed in São Paulo and Paraná. Avoador is a popular snack and side dish for breakfast (Fig. 7.20) (Sampaio and Menezes 2020).

Both sour and sweet starch can be used to prepare avoador. For 500 g of starch, mix with 250 mL of milk, 250 mL of water, 250 mL of oil, 1 egg, and 1 teaspoon of salt.

Preheat the oven to low temperature before starting to prepare the recipe. Mix the egg and salt in a bowl. Then add warm oil and milk. Gradually add the starch and stir until you get a consistent dough. Put 250 mL of boiling water into the mix and combine all the ingredients.

Put the dough in a pastry bag (or plastic bag) with a spout. Grease a pan with margarine and make the cookies in the shape you want, but do not let the cookies touch each other while baking. The cookies should be baked at 180°C for 20 minutes.

7.3.2.5 Pão de Queijo (Cheese Bread)

Pão de queijo, or Brazilian cheese bread, is a small, baked cheese roll or cheese bun, and a popular snack and breakfast food in Brazil. It is a traditional Brazilian recipe, originating in the state of Minas Gerais.

Pão de queijo consists of a type of biscuit with sour or sweet *polvilho* plus eggs, salt, vegetable oil, and cheese, of soft and elastic consistency (Fig. 7.21). There are several different recipes that vary the ingredients and type of cheese to be used. Natural and/or sour starches can be used to prepare the bread.

For 500 g of starch, add 300 mL of milk, 150 mL of oil, 3 eggs, 250 g of grated *Minas* cheese, and a teaspoon of salt.

In a bowl, heat the salt, milk, and oil. When boiling, scald the starch with this hot mixture, stir well, and let it cool. Add the eggs one by one, alternating with the cheese and kneading well after each addition. Grease your hands with oil if necessary.

Mold cookies 3–5 cm in diameter and place them on a greased baking sheet. Take to a preheated oven (180 °C) and bake until golden brown (around 15 minutes).

Fig. 7.21 Traditional Brazilian *pão de queijo* (cheese bread). (Photo: Podium Alimentos)

7.4 Closing Remarks

All the technological innovations detailed here are characterized by a high level of technological maturity, which means they are qualified technologies, fully adopted in Brazil, and, potentially, ready to be transferred directly, or with small adaptations, to other international production systems. On the other hand, their positive impacts on society do not seem to have been properly analyzed to create a vision of their success and how these technologies can be framed within the readiness scale recently developed by CGIAR (Sartas et al. 2020).

It is hoped these technology transfer trainings will allow participants to apply the skills in their own contexts. Therefore, it is important to design a training program that is sustainable and replicable in other regions or countries.

Considering this and based on the experiences acquired over the last 15 years, *Embrapa Mandioca e Fruticultura* has been trying to improve the implementation of international training to maximize the returns on investment.

Appropriate selection of participants is crucial. Typically, training candidates are chosen directly by the beneficiary institution, following the view that innovations can be transferred simply by intermediaries and change agents (e.g., extension workers) and then spread throughout the communities of individual beneficiaries. This conception has been widely refuted (Sartas et al. 2020).

In the first training sessions carried out by Embrapa, the African candidates selected were predominantly adult professionals, based in national R&D institutions primarily focused on administrative management activities rather than research. Usually, these individuals held administrative positions; some were not even involved with cassava crop. This lack of proximity to cassava production may have hindered their effectiveness as a technology multiplier inside and outside their institutions.

In the more recent cassava courses carried out by Embrapa, international organizing institutions (usually CGIAR and African NARIs), training sponsors, and funding agencies have selected young candidates with an entrepreneur profile, technical professionals, or recently graduated individuals linked to state institutions, private companies, and/or start-ups. Most young trainees consider training as essential to financial survival, an opportunity for them to introduce innovations in their own business and add value to their products.

Embrapa has specified how its trainings could be adapted or suited to address gender differences; however, we did not see the need to adjust the trainings to meet the preferences of men and women. The instructors – who are women and men – have noticed some differences in participation. Women showed more interest in topics related to techniques for preparing products from cassava processing. Men expressed greater interest in field activities. As the trainings occasionally require overnight stays or long hours away from home, this may create some implications that deserve attention to meet the needs and preferences of both men and women participants.

Participants of the three courses held within the Youth Technical Training Program on Cassava Post-harvest Processing (2017–2019) are already applying what they have learned in Brazil, and many of them provide to IBRAF and Embrapa with great feedback on the results of the Program at the micro level. One of the participants who founded the PaaClee Cassava Processing Company in 2014, for example, started to invest in training sessions with his employees, sharing the knowledge acquired during the 1-week course in Brazil. The YTTP on Post-harvest Processing of Cassava success also resulted in the organization of four YTTP sessions on Smallholder Agriculture with Senegalese learners and one YTTP session on Civic Engagement with Angolan learners, the first ones to be organized in bilateral arrangements, scheduled for 2021 and 2022 (UNDP-IBRAF 2020). During the Fourth International Cassava Conference (GCP21-IV), held in Benin in 2018, the young woman from Benin, *Paula Gnancadja*, a Bachelor in Management of Agricultural and Rural Enterprises, who participated in the YTTP training the previous year, presented the work "A Young Cassava Farmer and Processor Building a New Start-Up in Benin," reporting the success of her start-up (ABB.Sarl), after returning from Brazil, introducing new cassava processing products, which are being very well accepted by Benin consumers. She received the GCP21-IV Best Poster Award in the "entrepreneurship" category.

The technologies that received the greatest interest in the trainings include those products made from starch (*polvilhos*), such as beijus, tapiocas, and biscuits (avoador), and the minimal processing technique as a means for storing roots and using it later as raw material for chips, cakes, and dough for snacks.

One of the great challenges for the implementation of training for Africans is the great cultural and language diversity, coupled with the fact that the majority of Brazilian instructors speak only Portuguese. At the three YTTP courses, most of the African attendees spoke either English or French, while the Brazilian instructors spoke Portuguese, which create some communication difficulties.

The fact that Brazil is the center of origin and domestication of cassava explains the great importance of this crop, as well as the enormous diversity of its use and processing in different regions of Brazil. This time lag and the great genetic diversity of cassava made Brazil a major source of technologies for processing, which would later be mechanized and scaled for industry. Currently, the African cassava-producing countries are investing heavily in both sectors (family and industrial) to develop their own new technologies and/or to adapt or improve Brazilian innovations according to their demands. These efforts are welcome for government development programs, because of the unquestionable importance of cassava to reduce hunger and increase the income of cassava producers and processors. Therefore, we believe that the next cassava technologies transfer initiatives between Brazil and Africa should be extended to technologies, services, and products aimed at large-scale cassava processing and cultivation.

Acknowledgments The authors are grateful for the invitation and support of the CGIAR Research Program on Roots, Tubers and Bananas (RTB) in carrying out this chapter. We also thank the agencies that funded the courses reported here, such as Japan International Cooperation Agency (JICA), International Fund for Agricultural Development (IFAD), and African Development Bank (AfDB), as well as researchers and technicians from Embrapa and collaborators, who acted as training instructors.

References

AfDB (2017) African Development Bank. Brazil to groom African youth in cassava processing. https://www.afdb.org/en/news-and-events/afdb-brazil-to-groom-african-youth-in-cassava-processing-17348. Accessed 5 Jan 2021

Alves-Pereira A, Clement CR, Picanço-Rodrigues D, Veasey EA, Dequigiovanni G, Ramos SLF, Pinheiro JB, Zucchi MI (2018) Patterns of nuclear and chloroplast genetic diversity and structure of manioc along major Brazilian Amazonian rivers. Ann Bot 121:625–639

Araujo FCB, Moura EF, Cunha RL, Farias Neto JT, Silva RS (2019) Chemical root traits differentiate 'bitter' and 'sweet' cassava accessions from the Amazon. Crop Breed Appl Biotechnol 19:77–85

Bechoff A, Tomlins KI, Chijioke U, Ilona P, Bennett B, Westby A, Boy E (2019) Variability in traditional processing of gari: a major food security product from cassava. Food Chain 8:39–57

Falade KO, Akingbala JO (2011) Utilization of cassava for food. Food Rev Intl 27:51–83

FAOSTAT (2018) Food and Agricultural Organization of the United Nations Online Statistics Database. http://www.fao.org/faostat/en/#data/FBS. Accessed 4 Jan 2021

FAOSTAT (2019) Food and Agricultural Organization of the United Nations Online Statistics Database. http://www.fao.org/faostat/en/#data/QC. Accessed 4 Jan 2021

Ferreira Filho JR, Silveira HF, Macedo JJG, Lima MB, Cardoso CEL (2013) Cultivo, processamento e uso da mandioca – Instruções práticas. *Embrapa Mandioca e Fruticultura*, Cruz das Almas-BA, 32p

Folegatti MIS, Matsuura FCAU, Ferreira Filho JR (2005) A indústria da farinha de mandioca. In: Souza LS, Farias ARN, Mattos PLP, Fukuda WMG (eds) Processamento e utilização da mandioca. Embrapa Mandioca e Fruticultura Tropical, Cruz das Almas-BA, pp 61–141

Montagnac JA, Davis CR, Tanumihardjo SA (2009) Processing techniques to reduce toxicity and antinutrients of cassava for use as a staple food. Compr Rev Food Sci Food Saf 8:17–27

Morante N, Sánchez T, Ceballos H, Calle F, Pérez JC, Egesi C, Cuambe CE, Escobar AF, Ortiz D, Chávez AL, Fregene M (2010) Tolerance to postharvest physiological deterioration in cassava roots. Crop Sci 50:1333

Motta JS (2012) Beijus coloridos preparados com frutas e hortaliças. Embrapa Mandioca e Fruticultura, Cruz das Almas-BA, 2 p

Ndjouenkeu R, Kegah FN, Teeken B, Okoye B, Madu T, Olaosebikan OD, Chijioke U, Bello A, Osunbade AO, Owoade D, Takam-Tchuente NH, Njeufa EB, Chomdom ILN, Forsythe L, Maziya-Dixon B, Fliedel G (2021) From cassava to gari: mapping of quality characteristics and end-user preferences in Cameroon and Nigeria. Int J Food Sci Technol 56:1223–1238

NRI (2017) Natural Resources Institute. https://www.nri.org/latest/news/2017/nri-signs-cooperation-agreement-with-embrapa-brazilian-agricultural-research-corporation. Accessed 8 Nov 2020

Oliveira LA, Motta JS, Jesus JL, Sasaki FFC, Viana ES (2020a) Processing of sweet and bitter cassava. Embrapa Mandioca e Fruticultura, Cruz das Almas-BA, 64 p

Oliveira LA, Reis RC, Viana ES, Santos JF, Souza VS, Assis JLJ, Sasaki FFC, Santos VS (2020b) Obtenção de raízes de mandioca de mesa congelada. Embrapa Mandioca e Fruticultura, Cruz das Almas-BA. 8p. (*Embrapa Mandioca e Fruticultura*. Comunicado Técnico, 177)

Oliveira LA, Godoy RCB (2011) Mandioca chips. Embrapa Mandioca e Fruticultura, Cruz das Almas-BA. 6p. (*Embrapa Mandioca e Fruticultura*. Circular Técnica, 101)

Sampaio VS, Menezes SSM (2020) A produção artesanal de biscoitos em Vitória da Conquista-

Bahia. Caderno Prudentino de Geografia 42:79–97
Sartas M, Schut M, Proietti C, Thiele G, Leeuwis C (2020) Scaling readiness: science and practice of an approach to enhance impact of research for development. Agric Syst 183:102874, 12 pp
SENAR (2018) Serviço Nacional de Aprendizagem Rural. Agroindústria: produção de derivados da mandioca. Coleção SENAR – 214. Brasília-DF: SENAR, 72 pp
Tomlins K, Bennett B (2017) New uses and processes for cassava. In: Hershey CH (ed) Achieving sustainable cultivation of cassava volume 1: cultivation techniques. Burleigh Dodds Science, Cambridge UK, pp 89–98
UNDP-IBRAF (2020) United Nations Development Programme – Brazil Africa Institute. Good practices on south-south cooperation in the context of food security. UNDP, New York. 90 pp
Viana ES, Oliveira LA, Silva J (2010) Processamento mínimo de mandioca. Embrapa Mandioca e Fruticultura, Cruz das Almas-BA. 4p. (*Embrapa Mandioca e Fruticultura.* Circular Técnica, 95)

8

Better Safety Norms for Cassava Products

Linley Chiwona-Karltun, Leon Brimer, and Jose Jackson

Abstract Cassava was domesticated in the Amazon Basin, where Native Americans selected many bitter varieties, and devised methods for detoxifying them. Cassava reached Africa in the sixteenth century, where rural people soon learned to remove the cyanogenic toxins, e.g., by drying and fermenting the roots. Processing cassava to remove the cyanogenic toxins including the cyanide formed during the processing is time consuming. The work is often done by women, while women and men often prefer bitter cassava varieties for social reasons and superior taste and color. In spite of deep, local knowledge of safe processing, traditional foods made with contaminated water may contain bacterial and fungal pathogens. Improper storage may encourage mycotoxins, such as aflatoxin. Recent advances in industrial processing are developing foods that are free of toxins and microbial contamination. Processing and selling cassava leaves is an emerging but fast-growing sector. Cassava leaves also contain cyanogenic toxins normally in higher concentrations than the cassava roots. In the future, more attention must be paid to the safe processing of cassava leaves and roots, especially as food processing becomes increasingly industrialized worldwide.

L. Chiwona-Karltun (✉)
Department of Urban & Rural Development, Swedish University of Agricultural Sciences, Uppsala, Sweden
e-mail: Linley.chiwona.karltun@slu.se

L. Brimer
Department of Veterinary and Animal Sciences University of Copenhagen, Frederiksberg, Denmark

J. Jackson
Alliance for African Partnership, International Studies and Programs, Michigan State University, East Lansing, MI, USA
e-mail: jacks184@msu.edu

8.1 Improving Safety of Cassava Products in Regional Cassava Production and Processing

The safety of a raw material or a processed food product can be classed into physical, microbiological, and chemical safety. Physical safety, such as sharp particles or contaminants in food, is not discussed here; the focus is on microbiological and chemical threats. The last decade has witnessed a rise in the quality of research and development on the management of pathogenic bacteria and mycotoxin-producing fungi in cassava products. Several research-funding agencies, at times in partnership with the CGIAR programs, have supported much of this development, discussed in this chapter, particularly regarding cassava food safety and human health.

A persistent challenge is how research and education integrate traditional indigenous knowledge on food toxicity, processing, preparation, and food safety into modern research on food technology. Many curricula fail to teach the role of traditional indigenous knowledge, science and technology, and their implications for Africa's food systems. Historically, cassava food safety evolved in the context of practices originating from cassava's region of origin in lowland South America, and elsewhere where the crop was taken since the sixteenth century often linked to slavery and trade. These interactions have laid the foundations for the repertoire of skills in processing, preparation, preferences for taste, cassava products, potential for modification, and consumer preferences.

The chemical safety of a plant-derived food may be affected by the environmental occurrence of toxic anthropogenic compounds. Furthermore, chemical safety is affected by toxic elements concentrated by the plant, e.g., selenium and cadmium, toxic secondary organic metabolites formed by the plant, and toxic organic compounds formed during food processing such as acrylamide.

This chapter focuses on the safety of foods derived from cassava roots and leaves, progress in research and recent consumer legislation worldwide. The chapter also addresses the toxic compounds the plant forms, cyanogenic toxins in the form of cyanogenic glycosides and their products of degradation, as well on aflatoxins that are carcinogenic mycotoxins. The chapter ends with a section on future needs and policy recommendations with emphasis on the opportunities emerging with the African Continental Free Trade Area, particularly for women.

8.2 Introduction

8.2.1 Traditional Indigenous Knowledge in Cassava

According to the World Intellectual Property Organization, "Traditional knowledge (TK) is knowledge, know-how, skills and practices that are developed, sustained and passed on from generation to generation within a community, often forming part of its cultural or spiritual identity" (WIPO 2021). Nowhere is this passing of traditional knowledge so closely interlinked as in the domestication and utilization of cassava as a food in the diet. Cassava is one of about 100 species of the genus *Manihot* that also includes several rubber-producing plants (Schery 1947). This genus is a member of the Euphorbiaceae family, and *Manihot esculenta* is the only plant in that family that produces tuberous roots that have led to its domestication (Schery 1947). Cassava was domesticated in the Amazon region purposively selected for its fleshy roots and wide range in color (white to yellow), starch and taste, particularly based on sugar content (Schaal et al. 2006). Amazonian communities have long interacted with cassava as evidenced by their diverse uses of its roots and leaves (Schwerin 1971). Unlike most other places that have adopted cassava, the whole cassava plant is used, not just the roots (Schaal et al. 2006). During domestication, cassava landraces with high and low levels of cyanogenic glycosides appeared. Toxic types were associated with lower population density and shifting agriculture with more isolated gardens where cyanogens provided protection against damage by animals. The Amazonian peoples developed sophisticated technology to detoxify the bitter types for consumption (McKey et al. 2010). Amazonian peoples have a deep understanding of cassava and of its uses (Carvalho et al. 2004). For example, sugary landraces were used for making natural sweetened drinks; amylose-free starch varieties were fed to young children as this was easier on the stomach; deep yellow colored cassava roots were used for making the traditional drink *tucupi*. Native Amazonian Tukanoan and Jivaroan peoples used cassava leaves as a vegetable, grated and boiled (Dufour 1994). Scientific publications, especially earlier ones, depict cassava as a food that is poor in nutrients, which should be discouraged as food for children. Scientists have learned and still can learn from indigenous communities with deep knowledge of the preparation and consumption of cassava. Such studies demand innovative methods, interdisciplinary teams combining scientists, practitioners, feminist food scholars, and those with a long history of farming and using cassava in Africa as well as worldwide.

As cassava became more and more domesticated and populations moved in search of more land or better opportunities, these agrarian-based families often carried the local and favorite cassava cuttings with them. The Carib- and Arawak-speaking peoples who migrated from South America took cassava with them to the Caribbean (Cousins 1903). For cassava to be readily adopted in Africa, there may have been some existing knowledge of processing toxic plants, such as of processing toxic yams on the west coast of Africa (Ohadike 1981). Because the slave ships often carried cassava as a provision during their travels between the New World and Africa, some processing methods like that of *gari* were brought over to Africa and adapted locally (Northrup and McCammon 1980). In her book on eating wild plants safely, Williams provides ten basic rules about how to enjoy the edible ones without endangering oneself with the poisonous ones. Rule #6 and rule #10 are particularly useful and remain relevant in light of the ensuing section on the chemical food safety of cassava products.

Box 8.1 Basic Rules for Eating Wild Plants Safely
Rule #6
"Know what part of the plant is edible and when it is edible."
Rule #10
"Know the common poisonous plants of the area." A good question to ask when learning any new plant is, "Is there anything poisonous that looks like this?"
Source: Kim Williams 1977. Eating Wild Plants. P 3–5.

8.3 Chemical Safety of Cassava Food Products – The Story About Cyanogens

Of the approximately 374,000 plant species currently described (Christenhusz and Byng 2016), at least 2500 contain cyanogenic glycosides (Zagrobelny et al. 2008). These compounds have important roles in plant defense due to their bitter taste and to their release of the highly toxic compound hydrogen cyanide (HCN) upon plant tissue disruption. Recent research points to additional roles as storage compounds of reduced nitrogen and sugar (Zagrobelny et al. 2008).

Cassava is one of a few plants that contain potentially toxic cyanogenic glyco-sides, yet are still used as food. Adolf Nahrstedt (1993) reviewed cyanogenesis and food plants, including cassava, lima bean, sorghum, bamboo, flax, bitter almond, apricot, cherry, and laurel cherry. An evaluation of the health risks of cyanogenic glycosides in foods (EFSA 2019) includes the seven mentioned on the Nahrstedt's list and no additional species. The roots and leaves of fresh cassava release hydrogen cyanide (HCN) during chewing and processing and storage. More than 291 million tons of cassava were produced worldwide in 2017, of which Africa accounted for over 60%. In 2017, Nigeria produced 59 million tons, making it the world's largest producer (about 20% of global production) with a 37% increase in the last decade (FAO 2017). The present status concerning production of safe food from cassava worldwide is discussed below.

8.4 Microbial Safety of Cassava Products (Bacteria, Fungi, and Mycotoxins)

Interest in the microflora associated with cassava roots and foods made from them started around the mid-1980s. Many of the root-derived foods are fermented, such as the sun-dried sour starch, which is mainly produced in Colombia and in Brazil.

Table 8.1 Microflora of cassava flour processed by domestic heap fermentation

Microflora	Mean ^{10}LogN/g (6 samples)
Total aerobic mesophiles	8.45
Lactobacillus spp.	7.50
Micrococcaceae	7.30
Pseudomonas spp.	6.60
Enterobacteriaceae	6.53
Bacterial endospores	6.30
Coliform bacteria	6.28
Escherichia coli	4.00
Bacillus cereus	<2.7
Clostridium perfringens	<1.7
Staphylococcus aureus	<1.7
Salmonella	Absent in 25 g
Yeasts	6.62
Filamentous fungi	5.00

Source: Based on Essers et al. (1995a)

The production of this and other fermented cassava root products involves a lactic acid fermentation and a decrease in pH down to around 3.5. However, the quality varies widely (Morlon-Guyot et al. 1998).

Studies of the fungal flora show that the mucoraceous fungus *Rhizopus oryzae*, associated with the postharvest spoilage of the roots, effectively degrades linamarin (a type of cyanogenic glucoside) (Padmaja and Balagopal 1985). By the end of the 1980s, certain cassava root products, such as flours, were included in Brazilian studies on the occurrence of mycotoxins in foods (Soarez and Rodrigues-Amaya 1989).

During the 1990s, studies began to report on the total microflora, i.e., species of bacteria, yeasts, and filamentous fungi (Essers et al. 1995a, b, c). In the study entitled "Reducing Cassava Toxicity by Heap-Fermentation in Uganda," Essers et al. (1995a) identified and quantified (10LogN/g flour) a number of species of bacteria, some bacterial groups, total yeasts, and total filamentous fungi in flour made from cassava roots that had undergone domestic heap fermentation (Table 8.1).

Fourteen filamentous fungi were identified to species (Table 8.2). Extracts of 25 flours and one sample of dried scraped-off molds were analyzed for mutagenicity and cytotoxicity using the Ames test with the tester strains *Salmonella typhimurium* TA100 and TA98, with and without the addition of S9 liver homogenate, and also for the presence of any aflatoxins. The aflatoxins were analyzed by High Performance Thin Layer Chromatography (HPTLC) with a detection limit of 2 microgram per kilogram (Essers et al. 1995a).

The authors of the above study concluded that except for *Neurospora sitophila* and *Geotrichum candidum*, all other fungi identified (Table 8.2) were able to produce mycotoxins under cultured conditions. The high number of *Escherichia coli* in

Table 8.2 Filamentous fungi occurring in cassava flour processed by domestic heap-fermentation

Fungi	Number of samples in which detected ($n = 6$)
Aspergillus fumigatus	3
Aspergillus niger	1
Aspergillus oryzae	4
Aspergillus parasiticus	1
Fusarium sporotrichioides	1
Geotrichum candidum	6+
Mucor circinelloides	2
Mucor racemosus	3
Neurospora sitophila	6+
Neurospora crassa	1
Penicillium frequentans	1
Penicillium waksmanii	1
Rhizopus oryzae	5+
Rhizopus stolonifer	4

Note: + Numerous in each positive sample
Source: Based on Essers et al. (1995a)

one sample was linked to unhygienic processing, while the spore-forming *Clostridium perfringens* and *Bacillus cereus*, found sporadically, might produce toxins after cooking ugali, a maize dish popular in Tanzania and Kenya, if the grain had been stored for a long time. However, finding no cytotoxicity, mutagenicity, nor any aflatoxins, it was concluded that such mycotoxins were not formed in quantities that are detrimental to public health (Essers et al. 1995a).

The study by Essers et al. (1995a) points to the threat of pathogenic bacteria from contaminated water. A study in Nigeria found that the bacterial and coliform counts of fermented cassava flour (kpor umilin) were reduced considerably if the product was made with potable water and additional washing (Tsav-wua et al. 2004; Inyang et al. 2006).

8.4.1 The Development After Essers et al.

From 2007 to 2019, research on the safety of cassava food products suddenly increased. Thirteen articles are summarized in Table 8.3.

Table 8.3 Summary of articles on microbiological food safety for cassava products (2007–2019)

Year	Country	Cassava product(s)	Potentially hazardous microorganism identified (y/n/ NA)	+Mycotoxin(s) identified (y/n/ NA)	Recommendations/ conclusions given (y/n)	++ Ref. number
2007	Nigeria	Gari, lafun, ogiri	Y	NA	N	1
2008	Benin	Casava chips	Y	N	N	2
2008	Nigeria	Fufu	Y	NA	Y (HACCP strategy recommended)	3
2009	Nigeria	Lafun, fufu, gari	N	NA	N	4
2010	Uganda	Casava chips	Y	Y (aflatoxins)	Y (drying methods recommended)	5
2010	Nigeria	Wet fufu	Y	NA	Y (GMP recommended)	6
2012	Nigeria	Gari	NA (only total counts analyzed)	NA	Y (type of packaging material recommended)	7
2012	Benin	Cassava chips	Y	N	Y (maize and groundnuts are higher in aflatoxins)	8
2013	Nigeria	Fermented cassava flour "lafun"	Y	NA	Y (improved water quality and health inspections during production)	9
2014	Nigeria	Gari	NA (only total counts analyzed)	NA	Y (keep moisture low by proper packaging)	10
2016	Kenya	Cassava chips, flour	Y (heavy contamination with coliforms and for one sample with *E. coli*)	NA	Y (proper training of producers)	11
2017	Nigeria	9 products ++	NA	Y (aflatoxins and fumonisins)	Y (mycotoxin levels are low)	12
2019	Nigeria	Cassava root (white and yellow) fermented 4 days	Y/N (*E. coli* found after 48 hours of fermentation, but not after 72 and 96 hours)	NA	N	13

(continued)

Table 8.3 (continued)

+NA (not analyzed for)

++References by numbers: 1, Ljabadeniyi (2007); 2, Gnonlonfin et al. (2008); 3, Obadina et al. (2008); 4, Obadina et al. (2009); 5, Kaaya and Eboku (2010); 6, Obabina et al. (2010); 7, Adejumo and Raji (2012); 8, Gnonlonfin et al. (2012a, b); 9, Adebayo-Oyetoro et al. (2013); 10, Olopade et al. (2014); 11, Gacheru et al. (2016); 12, Abass et al. (2017); 13, Obi and Ugwu (2019); 14, Odu and Maduka (2019)

+++Cassava starch, HQCF, lafun, fufu flour, tapioca, fine yellow gari, fine white gari, yellow kpo-kpo gari, white kpo-kpo gari

Note: This microbiological research dealt with the most analyzed commercialized products bought at selected markets or supermarkets, while a few of the studies included products produced by the scientists themselves in the laboratory. The analytical methods may have different limits of detection or of quantification for mycotoxins

8.4.2 The Developments Made Possible

One pleasant surprise from this research was the absence or exceptionally low level of aflatoxins in some of the studies, some of which even failed to find the fungus *Aspergillus flavus*, which causes aflatoxins (Abass et al. 2017; Kaaya and Eboku 2010; Gnonlonfin et al. 2012a, b), especially considering that these studies did use advanced analytical methods. During this period, several manuals were published, such as *Potential for Commercial Production and Marketing of Cassava: Experiences from the Small-Scale Cassava Processing Project in East and Southern Africa* (Abass et al. 2013), *Quality Management Manual for Production of High Quality Cassava Flour* (Dziedzoave et al. 2006), and *Quality Management Manual for the Production of Gari* (Abass et al. 2012), suggesting improving conditions for developing commercial cassava food products on a larger scale. Such more advanced business developments even can find further support by the fact that an expert group with representatives from (1) University of Natural Resources and Life Sciences in Vienna, Austria, (2) Kwara State University in Nigeria, (3) International Fund for Agricultural Development, and (4) IITA in 2017 published the extremely important paper "Assessment of the Potential Industrial Applications of Commercial Dried Cassava Products in Nigeria" (Awoyale et al. 2017). In this paper, they summarized seven functional properties and seven pasting properties for 13 cassava product groups, based on samples collected from cassava processors and marketers.

8.4.3 The Cyanogenesis of Cassava

The major cyanogenic compound in cassava is linamarin. As it degrades, it releases the toxic HCN, through a two step process. That is the reactions with a glucosidase (linamarase) followed by a reaction with a second enzyme (hydroxynitrile lyase). Further in-depth details are shown in the risk assessment by the European Food Safety Authority (EFSA 2019, p. 11). The reactions start when the cyanogenic glucoside and the enzymes are brought in contact through rupture of the plant tissues, in which the three components have been stored separately (Behera and Ray 2016). Intact cassava roots may release less than 10 mg HCN/kg fresh tissue or as much as 1100 (EFSA 2019).

In 2003, Codex Alimentarius Commission released the Codex Standard for Sweet Cassava (CODEX STAN 238-2003), which has been amended three times, i.e., 2005, 2011, and 2013. This standard among others regulates the maximum content of cyanogenic glucosides.

Box 8.2 Codex Standard for Sweet Cassava (CODEX STAN 238-2003)
"Sweet varieties of cassava are those that contain less than 50 mg/kg hydrogen cyanide (fresh weight basis). In any case, cassava must be peeled and fully cooked before being consumed."

In 2010, the 238-standard was followed by another standard, this time for bitter cassava root.

Box 8.3 CODEX STANDARD FOR BITTER CASSAVA1 (CODEX STAN 300-2010)
"This Standard applies to commercial bitter varieties of cassava roots grown from *Manihot esculenta* Crantz, of the Euphorbiaceae family, to be supplied fresh to the consumer, after preparation and packaging. Cassava for industrial processing is excluded. Bitter varieties of cassava are those containing more than 50 mg/kg of cyanides expressed as hydrogen cyanide (fresh weight basis). In addition to the requirements of the General Standard for the Labelling of Prepackaged Foods (CODEX STAN 1-1985), the following specific provisions apply: Each package shall be labelled as to the name of the produce and type (bitter) and may be labelled as to the name of the variety. A statement indicating the following is required: cassava must not be eaten raw; cassava shall be peeled, de-pithed, cut into pieces, rinsed and fully cooked before consumption; and cooking or rinsing water must not be consumed or used for other food preparation purposes."

While the commercial marketing of bitter cassava according to the CAC standard (CODEX STAN 300-2010) is limited, studies in Denmark (Kolind-Hansen and Brimer 2010), Ireland (O'Brien et al. 2013), and Australia (Burns et al. 2012) demonstrated many cassava roots sold as sweet were in fact bitter. Cassava safety becomes important as African migrants, 22 million in Europe alone, seek to consume foods from their home continent.

The toxicity of a cyanogenic food plant such as cassava has to do with the amount of HCN and cyanohydrins formed before or while eating. This is because the intact glycosides (e.g., linamarin) are stable in the digestive tract, but the cyanohydrins will split into HCN and its constituent aldehyde/ketone. The non-hydrolyzed part of the linamarin is absorbed and excreted unchanged in the urine (Brimer and Rosling 1993). The primary mode of action for acute toxicity of HCN is the inhibition of the cell's oxidative phosphorylation. This leads to anaerobic energy production. Due to their high oxygen and energy demand, the brain and heart are among the most sensitive organs. Thus, cyanide can result in hypoxia, metabolic acidosis, and impairment of several vital functions. The CONTAM Panel of EFSA in 2019 concluded that there are no data indicating what the acute reference dose (ARfD) for cyanide of 20 µg/kg bw, established in 2016, should be (EFSA 2019). Certain neurological disorders, e.g., spastic paraparesis (konzo) and tropical ataxic neuropathy, have been associated with chronic dietary exposure to cyanide. However, this has always been in populations where cassava was the main food source. In areas with low iodine intake, chronic cyanide exposure from cassava has also been associated with hypothyroidism and goiter. In 2012, the JECFA concluded that the epidemiological association between cassava consumption and konzo was consistent, even though the etiological mechanism of konzo is still unknown (cited in EFSA 2019).

Clearly, the content of cyanogenic glycosides in raw cassava roots and the effects of food processing on these toxins are of major importance. Below, we will take a further look.

8.4.4 *Traditional Processing: Reasons and the Products*

According to Olsen and Schaal (1999), wild populations of *M. esculenta* occur primarily in west central Brazil and eastern Peru. This region is also the home for Native Americans, including the Tukanoans, where Dufour studied 14 cassava varieties and their cyanogenic potentials (Dufour 1988). The mean potentials varied from around 310–561 mg HCN equivalents/kg f.w. for the *Kii* (= toxic cultivars) and down to around 170 for *Makasera* (nontoxic/safe), the only nontoxic cultivar grown. The toxic varieties were preferred for growing (Dufour 1988). These values can be compared with the accepted highest content of cyanogenic glycosides in sweet cassava roots as accepted by the CAC standard, i.e., 50.

Ten years later after Dufour's publication, a study by Chiwona-Karltun et al. (1998) found similar results in Malawi, where cassava had been brought by the Portuguese starting in the 1500s (Leestma 2015). The study found that bitter cultivars were often preferred, because of experience and traditional indigenous knowledge in processing, managing theft, and most of all ensuring food security (Chiwona-Karltun 2001; Chiwona-Karltun et al. 2000). Toxicity was not seen as a problem, although such roots needed processing for detoxification (Chiwona-Karltun et al. 2000) and the women were extremely adept at correlating bitter taste with toxicity (Chiwona-Karltun et al. 2004). In 1986, Fresco wrote: "throughout Africa, bitter varieties are much more common than sweet ones, notwithstanding the fact that sweet varieties have been promoted by some colonial administrations out of concern with the toxicity of bitter varieties. There is no satisfactory explanation in the literature for the predominance of bitter varieties" (Fresco 1986, p. 153). From the work in Malawi and studies elsewhere, it is clear that bitter varieties where they constitute a major part of the diet are socially preferred, notwithstanding the amount of work that is required and mostly undertaken by women, to process and render them safe for consumption.

8.5 Gender-Preferred Traits and Traditional Indigenous Foods

Now we understand that a considerable part of cassava roots grown worldwide needs to be processed in order to remove a major part of the cyanogenic glycosides present, thereby making a safe product, and in addition obtain one of several kinds of foods. Long processing times leave women with little time for other activities (Nweke 1999). This could have profound effects on the dietary diversity of a household as observed in the Amazon Basin (Dufour 1994) where women spent much time processing cassava. Processing is essential to reduce the water content and preserve the very perishable roots in a way they can be more easily transported.

Processing is also related to gendered variety and trait preference. In the works published by Chiwona et al. in (1998), the reasons for the preferences for bitter varieties were linked with experience and possession of traditional indigenous knowledge in processing, managing social changes in the society such as theft, and most of all ensuring food security. The studies also revealed that bitter cassava cultivars had certain traits that provided products that were deemed to be of superior quality, e.g., processed cassava flour is perceived to be finer and whiter in the making of the staple dish *kondowole* (Chiwona-Karltun et al. 1998). Cassava leaf foods from bitter varieties are regarded as tastier, also verified in Zambia (Chiwona-Karltun et al. 2015). Most importantly, bitter cassava reduced the fear of being without food, especially for women that were "vulnerable in society." These women were usually in female-headed households, and they needed to protect their food sources based on the deep knowledge of food processing and preparation. Incorporating gender-preferred traits continues to be a challenge in breeding programs (Weltzien et al. 2019) especially when these priorities compete with urban demand for food, changing diets and preferences for processed foods (Raheem & Chukwuma 2001).

It is worthwhile reflecting on how to preserve traditional food knowledge while understanding these preferred cassava traits from a gendered perspective. For example, in Jamaica, the popularity of *bammy bread* (Fig. 8.1), a preferred traditional food has been preserved largely through the women to women transfer of knowledge and expertise. By way of the kitchen and cuisine identity, bammy bread has successfully been incorporated in modern bakeries and extended to beyond export markets (Canevari-Luzardo et al. 2020). As climate change dampened the production and availability of cassava, Jamaica stopped exporting bammy and only served the national market. This calls for paying closer attention to the roles that women and men play in cassava value chains (Andersson et al. 2016), including how specific products impact women and men and their well-being. As traditional foods are

Fig. 8.1 Traditional fried bammy bread. Photo credit Sam Biehl

Fig. 8.2 Mortar used for pounding cassava leaves (left) (Photo credit Brian LeniKanjeri Msukwa, Malawi). Preferred locally pounded cassava leaves being cooked (right). (Photo credit D. Dufour (CIRAD))

increasingly appreciated for their contribution to health and well-being, demand for locally made and processed foods will gain more traction.

Cassava leaves also play important roles. Fresh or dried cassava leaves are rarely sold on the market, largely due to the focus on cassava roots as an energy provider. But now cassava leaf production and marketing are well established in the Democratic Republic of the Congo, Rwanda, Angola, and Malawi to mention a few examples and have even reached, e.g., European markets in frozen form as sold in special shops. The technology for processing, packaging, and marketing remains a challenge. To render cassava leaves safe to eat, they must be disintegrated in order for the endogenous enzyme to initiate the breakdown of the cyanogenic glucosides (Bokanga 1994). Pounding for at least 15–20 minutes is crucial. Few affordable leaf-processing machines are available (Fig. 8.2).

8.5.1 *The Traditional Processing Reasons and the Products*

Several cassava root products were developed in the center or origin of the crop. The manual *Cassava Products of Brazil* lists starch, sour starch, *puba* flour (similar to African *fufu*), fermented roasted flour (similar to gari), *tucupi* (fermented savory syrup), tiquira, unfermented gari (*farinha de mandioca*), and cassava crisps (Crenn 2019). Various scientific articles and book chapters published after 2010 have described the different unit processes and their impact on the composition of cassava root-based food products (Gnonlonfin and Brimer 2013; Gnonlonfin et al. 2012a, b). Important processes include drying, soaking, cooking (baking, frying, steaming), size reduction (grating), fermentation (dry in heap or wet), and smoking (to make the smoked balls called *kumkum*).

Fermentation can be used to make cassava products that are free of cyanogenic toxins, as discussed in the following section.

8.5.2 *The Progress in Knowledge and Technology*

Gari is the most well-known fermented cassava food product found throughout much of West Africa. Today, it comes in several professional and internationally marketed brands. Yet in many places, gari is still produced locally to eat at home or to sell. The CGIAR organizations have been involved in training and research on gari.

Starting with the manual *Quality Management Manual for the Production of Gari* (Abbas 2012) and following up with workshops like "Nigeria's Gari Revolution: Improving Efficiency and Equity of a Staple Food" (organized by Global Cassava Partnership for the 21st Century in 2016), this important fermented product is becoming professionalized. However, depending on the naturally occurring microorganisms for its fermentation, gari and other fermented cassava products still have room for improvement. The possibilities include (1) the selection and characterization of microorganisms found in safe products and (2) microorganisms which upon testing may be candidates for more complete degradation of the cyanogenic glycosides in cassava roots.

Penido et al. (2018) selected and characterized microorganisms (lactic acid bacteria and yeasts) originating from the production of Brazilian sour cassava starch, while Kostinek et al. (2007) isolated 375 predominantly lactic acid bacteria from fermenting cassava in different African countries. *Lactobacillus plantarum* isolated from fermenting cassava can hydrolyze (degrade) cyanogenic glycosides (e.g., linamarin), as can different *Bacillus subtilis* isolates collected from low pH starters used to produce *gergoush* (a Sudanese fermented snack) (Lei et al. 1999; Abban et al. 2013). See also the chapter by Behera and Ray (2016) on "Microbial Linamarase in Cassava Fermentation."

Fig. 8.3 Grating sweet cassava by hand (left), (Photo credit Joyce NaMtawali Mwagomba, Malawi) and sun drying the grated cassava on mats to produce high-quality cassava flour (HQCF, right). (Photo credit D. Dufour (CIRAD))

8.6 Conclusions and Looking Forward

Traditional indigenous cassava-based food products have been developing greater food safety. Worldwide, more and more popular drinks like *bubble* or *boba* tea with tapioca "pearls" have been developed with very low levels of cyanogens (Bulathgama et al. 2020). New food products based on cassava must focus on these toxins to develop safe products.

As more women venture into value addition especially at local scale and hoping to capitalize from the African Continental Free Trade Area and the opportunity it may accord women, we need to enable women's participation. However, there are some immediate challenges for women and men, such as lack of access and skills to resources to scale-up production. For example, large-scale processing of high-quality cassava flour or cassava flour and cassava leaves is still rudimentary. Packaging and meeting standards for export requires training and investments that often are beyond the reach of women or small-scale farmers (Fig. 8.3).

References

Abass AB, Dziedzoave NT, Alenkhe BE, James BD (2012) Quality management manual for the production of gari. IITA, Ibadan, Nigeria.

Abass A, Mlingi N, Ranaivoson R, Zulu M, Mukuka I, Abele S, Bachwenkizi B, Cromme N (2013) Potential for commercial production and marketing of cassava: experiences from the small-scale cassava processing project in East and Southern Africa. IITA, Ibadan, pp 13–88

Abass AB, Awoyale W, Sulyok M, Alamu EO (2017) Occurrence of regulated mycotoxins and other microbial metabolites in dried cassava products from Nigeria. Toxins 9(7):207. https://doi.org/10.3390/toxins9070207

Abban S, Brimer L, Abdelgadir WS, Jacobsen M, Thorsen L (2013) Screening for Bacillus subtilis group isolates that degrade cyanogens at pH 4.5–5.0. Int J Food Microbiol 161:31–35

Adebayo-Oyetoro AO, Oyewole OB, Obadina AO, Omemu MA (2013) Microbiological safety assessment of fermented cassava flour "Lafun" available in Ogun and Oyo states of Nigeria. Int I Food Sci 213: Article ID 845324, 5 pages. https://doi.org/10.1155/2013/845324

Adejumo BA, Raji AO (2012) Microbial safety and sensory attributes of gari in selected packaging materials. Acad Res Int 3(3):153–162

Andersson K, Lodin JB, Chiwona-Karltun L (2016) Gender dynamics in cassava leaves value chains: the case of Tanzania. J Gender Agric Food Security 1(2):84–109

Awoyale W, Abass AB, Ndavi M, Maziya-Dixon B, Sulyok M (2017) Assessment of the potential industrial applications of commercial dried cassava products in Nigeria. Food Measure 11:598–609

Behera SS, Ray RC (2016) Microbial linamarase in cassava fermentation. In: Ray RC, Rossel CM (eds) Microbial enzyme technology in food applications. Food biology series. CRC Press/Taylor & Frances Group, Boca Raton

Bokanga M (1994) Processing of cassava leaves for human consumption. Acta Hortic 375:203–208

Brimer L, Rosling H (1993) Microdiffusion method with solid state detection of cyanogenic glycosides from cassava in human urine. Food Chem Toxicol 31(8):599–603

Bulathgama BEAU, Gunasekara GDM, Wickramasinghe I, Somendrika MAD (2020) Development of commercial tapioca pearls used in bubble tea by microwave heat–moisture treatment in cassava starch modification. EJERS (Eur J Eng Res Sci) 5(1):103–106. https://doi.org/10.24018/ejers.2020.5.1.1455

Burns AE, Bradbury JH, Cavagnaro TR, Gleadow RM (2012) Total cyanide content of cassava food products in Australia. J Food Compos Anal 25(1):79–82

Canevari-Luzardo LM, Berkhout F, Pelling M (2020) A relational view of climate adaptation in the private sector: how do value chain interactions shape business perceptions of climate risk and adaptive behaviours? Bus Strateg Environ 29(2):432–444

Carvalho LJCB, de Souza CRB, de Mattos Cascardo JC, Junior CB, Campos L (2004) Identification and characterization of a novel cassava (Manihot esculenta Crantz) clone with high free sugar content and novel starch. Plant Mol Biol 56:643–659

Chiwona-Karltun L (2001) A reason to be bitter : cassava classification from the farmers' perspective. Institutionen för folkhälsovetenskap/Department of Public Health Sciences, Dissertation. Karolinska Institutet. Sweden.

Chiwona-Karltun L, Mkumbira JS, Saka J, Bovin M, Mahungu NM, Rosling H (1998) The importance of being bitter—a qualitative study on cassava cultivar preference in Malawi. Ecol Food Nutr 37(3):219–245

Chiwona-Karltun L, Tylleskar T, Mkumbira J, Gebre-Medhin M, Rosling H (2000) Low dietary cyanogen exposure from frequent consumption of potentially toxic cassava in Malawi. Int J Food Sci Nutr 51(1):33–43

Chiwona-Karltun L, Brimer L, Kalenga Saka JD, Mhone AR, Mkunbira J, Johansson L, Bokanga M, Mahungu NM, Rosling H (2004) Bitter taste in cassava roots correlates with cyanogenic glucoside levels. J Sci Food Agric 84:581–590

Chiwona-Karltun L, Nyirenda D, Mwansa CN, Kongor JE, Brimer L, Haggblade S, Afoakwa EO (2015) Farmer preference, utilization, and biochemical composition of improved cassava (Manihot esculenta Crantz) varieties in southeastern Africa. Econ Bot 69(1):1–15

Christenhusz MJM, Byng JW (2016) The number of known plants species in the world and its annual increase. Phytotaxa 261(3):201–217. https://doi.org/10.17660/ActaHortic.1994.375.18

Cousins HH (1903) Analysis of 17 varieties introduced from Colombia and grown by Mr. Robert Thompson at Half-Way-Tree. Jamaica Bull Dept Agric Jam 1(2):35

Crenn (2019) Cassava products of Brazil, IFAD and Natural Resources Institute. http://projects.nri.org/cassava-ipci/images/documents/Cassava_products_of_Brazil_a_production_guide_for_Africa.pdf

Dufour DL (1988) Cyanide content of cassava (Manihot esculenta, Euphorbiaceae) cultivars used by Tukanoan Indians in Northwest Amazonia. Econ Bot 42:255–266

Dufour DL (1994) Cassava in Amazonia: lessons in utilization and safety from native peoples. International Workshop on Cassava Safety. Acta Hortic 375:175–182. https://doi.org/10.17660/ActaHortic.1994.375.15

Dziedzoave NT, Abass AB, Amoa-Awua WK, Sablah M, Adegoke GO (2006) Quality management manual for production of high quality cassava flour. International Institute of Tropical Agriculture (IITA), Ibadan, p 68

EFSA (2019) Evaluation of the health risks related to the presence of cyanogenic glycosides in foods other than raw apricot kernels. EFSA J 17(4):5662–5740

Essers AJA, Ebong C, van der Grift RM, Nout MJR, Otim-Napa W, Rosling H (1995a) Reducing cassava toxicity by heap-fermentation in Uganda. Int J Food Sci Nutr 46(2):125–136

Essers AJA, Jurgens CMGA, Nout MJR (1995b) Contribution of selected fungi to the reduction of cyanogen levels during solid substrate fermentation of cassava. Int J Food Microbiol 26(2):251–257

Essers AJA, Bennik MHJ, Nout MJR (1995c) Mechanisms of increased linamarin degradation during solid-substrate fermentation of cassava. World J Microbiol Biotechnol 11:266–270

FAO (2017) Food outlook – Food and Agriculture Organization of the United Nations. Biannual report on global food markets November 2017

Fresco LO (1986) Cassava in shifting cultivation: a systems approach to agricultural technology development in Africa. Royal Tropical Institute, Amsterdam

Gacheru PK, Abong GO, Okoth MW, Lamuka PO, Shibairo SA, Katama CKM (2016) Microbiological safety and quality of dried cassava chips and flour sold in the Nairobi and coastal regions of Kenya. Afr Crop Sci J 24(suppl sI):137–143

Gnonlonfin GJB, Brimer L (2013) Cassava (Manihot esculenta Crantz) as a source of chemically safe food: a critical review. In: Chloe M, Jones DB (eds) Processed foods. Quality, safety characteristics and health implications. Nova Publishers, New York, pp 59–81

Gnonlonfin GJB, Hell K, Fandohan P, Siame AB (2008) Mycoflora and natural occurrence of aflatoxins and fumonisin B1 in cassava and yam chips from Benin. West Afr Int J Food Microbiol 122:140–147

Gnonlonfin GJ, Sanni A, Brimer L (2012a) Preservation of cassava (Manihot esculenta Crantz): a major crop to nourish people worldwide. In: Bhat R, Alias AK, Paliyath G (eds) Progress in food preservation. Wiley-Blackwell, Chichester, pp 331–342

Gnonlonfin GJB, Adjovi CSY, Katerere DR, Shephard GS, Sanni A, Brimer L (2012b) Mycoflora and absence of aflatoxin contamination of commercialized cassava chips in Benin, West Africa. Food Control 23:333–337

Inyang CU, Tsav-Wua JA, Akpapunam MA (2006) Impact of traditional processing methods on some physico chemical and sensory qualities of fermented cassava flour "Kpor umilin". Afr J Biotechnol 5(20):1985–1988

Kaaya A, Eboku D (2010) Mould and aflatoxin contamination of dried cassava chips in eastern Uganda: association with traditional processing and storage practices. J Biol Sci. https://doi.org/10.3923/jbs.2010.718.729

Kolind-Hansen L, Brimer L (2010) The retail market for fresh cassava root tubers in the European Union (EU): the case of Copenhagen, Denmark — a chemical food safety issue? J Sci Food Agric 90(2):252–256

Kostinek L, Sprecht VA, Edward VA, Pinto C, Egounlety M, Sossa C, Mbugua S, Dortu C, Thonart P, Taljaard M, Mengu C, Franz CMAP, Holzaplel WH (2007) Characterisation and biochemical properties of predominant lactic acid bacteria from fermenting cassava for selection as starter cultures. Int J Food Microbiol 114(3):342–351

Leestma J (2015) A tropical flour: Manioc in the Afro-Brazilian world, 1500–1800. https://digitalrepository.unm.edu/

Lei V, Amoa WKAA, Brimer L (1999) Degradation of cyanogenic glycosides by lactobacillus plantarum strains from spontaneous cassava fermentation and other microorganisms. Int J Food Microbiol 53(2–3):169–184

Ljabadeniyi AO (2007) Microbiological safety of gari, lafun and ogiri in Akure metropolis. Nigeria Afr J Biotechnol 6(22):2633–2635

McKey D, Cavagnaro TR, Cliff J, Gleadow R (2010) Chemical ecology in coupled human and natural systems: people, manioc, multitrophic interactions and global change. Chemoecology 20:109–133. https://doi.org/10.1007/s00049-010-0047-1

Morlon-Guyot J, Guyot JP, Pot B, de Haut IJ, Raimbault M (1998) Lactobacillus manihotivorans sp., a new starch-hydrolysing lactic acid bacterium isolated during cassava sour starch fermentation. Int J Syst Bacteriol 48:1101–1109

Nahrstedt A (1993) Cyanogenesis and foodplants. In: van Beek TA, Bretelser H (eds) Proceedings of the phytochemical society of Europe, Phytochemistry and Agriculture (Chapter 7). Clarendon Press, Oxford, pp 106–129

Northrup SH, McCammon JA (1980) Simulation methods for protein structure fluctuations. Biopolymers 19(5):1001–1016. https://doi.org/10.1002/bip.1980.360190506

Nweke FI (1999) Gender surprises in food production, processing, and marketing with emphasis on cassava in Africa (No. 19). IITA

O'Brien BGM, Weir RR, Moody K, Liu PWS (2013) Cyanogenic potential of fresh and frozen cassava on retail sale in three Irish cities: a snapshot survey. Int J Food Sci Technol 48(9):1815–1831

Obadina AO, Oyewole OB, Sanni LO, Tomlins KI, Westby A (2008) Identification of hazards and critical control points (CCP) for cassava fufu processing in South-West Nigeria. Food Control 19(1):22–26

Obadina AO, Oyewole OB, Odusami AO (2009) Microbial safety and quality assessment of some fermented cassava products (lafun, fufu, gari). Sci Res Essay 4(5):432–425

Obadina AO, Oyewole OB, Sanni LO, Tomlins KI, Westby A (2010) Improvement of the hygienic quality of wet "fufu" produced in South West Nigeria. Food Control 21:639–643

Obi CN, Ugwu CJ (2019) Effects of microbial fermentation on cyanide contents and proximate composition of cassava tubers. Nigerian J Biol 33(2):4493–4502

Odu NN, Maduka N (2019) Microbial analysis and molecular characterization of bacterial and fungal isolates present in exposed and packaged cassava, plantain and yam flour sold in selected markets in Port Harcourt, Rivers State, Nigeria. Am J Microbiol Res 7(2):63–72

Ohadike DC (1981) The influenza pandemic of 1918-19 and the spread of cassava cultivation on the lower Niger: a study in historical linkages. J Afr Hist 22(3):380–391

Olopade BK, Oranusi S, Ajala R, Olorunsola SJ (2014) Microbiological quality of fermented cassava (gari) sold in Ota Ogun state Nigeria. Int J Curr Microbiol App Sci 3(3):888–895

Olsen KM, Schaal BA (1999) Evidence on the origin of cassava: Phylogeography of Manihot esculenta. Proc Natl Acad Sci U S A 96(10):5586–5591

Padmaja G, Balagopal C (1985) Cyanide degradation by Rhizopus oryzae. Can J Microbiol 31:663–669

Penido FCL, Piló FB, Sandes SHDC, Nunes ÁC, Colen G, Oliveira, EDS, ... & Lacerda, ICA (2018) Selection of starter cultures for the production of sour cassava starch in a pilot-scale fermentation process. Braz J Microbiol 49:823–831

Raheem D, Chukwuma C (2001) Foods from cassava and their relevance to Nigeria and other African countries. Agric Hum Values 18:383–390

Schaal BA, Olsen KM, Carvalho LJCB (2006) Evolution, domestication, and agrobiodiversity in the tropical crop cassava. In: Darwin's harvest: new approaches to the origins, evolution, and conservation of crops, pp 269–284

Schery R (1947) Manioc-a tropical staff of life. Econ Bot 1(1):20–25. https://www.jstor.org/stable/4251837

Schwerin KH (1971) The bitter and the sweet, some implications of techniques for preparing manioc. In: Unpublished paper presented at the 1971 Annual Meeting of the American Anthropological Association

Soarez LMV, Rodrigues-Amaya DB (1989) Survey of aflatoxins, ochratoxin A, zearalenone, and sterigmatocystin in some Brazilian foods by using multi-toxin thin-layer chromatographic method. J Assoc Off Anal Chem 72(1):22–26

Tsav-wua JA, Inyang CU, Akpapunam MA (2004) Microbiological quality of fermented cassava flour "kpor umilin". Int J Food Sci Nutr 55(4):317–324

Weltzien E, Rattunde F, Christinck A, Isaacs K, Ashby J (2019) Gender and farmer preferences for varietal traits: evidence and issues for crop improvement. Plant Breed Rev 43:243–278

Williams K (1977) Eating wild plants. Mountain Press Pub

WIPO. Accessed 2021-05-24. https://www.wipo.int/tk/en/tk/

Zagrobelny M, Bak S, Møller BL (2008) Cyanogenesis in plants and arthropods. Phytochemistry 69:1457–1468

Permissions

All chapters in this book were first published by Springer; hereby published with permission under the Creative Commons Attribution License or equivalent. Every chapter published in this book has been scrutinized by our experts. Their significance has been extensively debated. The topics covered herein carry significant information for a comprehensive understanding. They may even be implemented as practical applications or may be referred to as a beginning point for further studies.

The contributors of this book come from diverse backgrounds, making this book a truly international effort. We would like to thank all the contributing authors for lending their expertise to make the book truly unique. They have played a crucial role in the development of this book. Without their invaluable contributions this book wouldn't have been possible. They have made vital efforts to compile up to date information on the varied aspects of this subject to make this book a valuable addition to the collection of many professionals and students.

This book was conceptualized with the vision of imparting up-to-date and integrated information in this field. To ensure the same, a matchless editorial board was set up. Every individual on the board went through rigorous rounds of assessment to prove their worth. After which they invested a large part of their time researching and compiling the most relevant data for our readers.

The editorial board has been involved in producing this book since its inception. They have spent rigorous hours researching and exploring the diverse topics which have resulted in the successful publishing of this book. They have passed on their knowledge of decades through this book. To expedite this challenging task, the publisher supported the team at every step. A small team of assistant editors was also appointed to further simplify the editing procedure and attain best results for the readers.

Apart from the editorial board, the designing team has also invested a significant amount of their time in understanding the subject and creating the most relevant covers. They scrutinized every image to scout for the most suitable representation of the subject and create an appropriate cover for the book.

The publishing team has been an ardent support to the editorial, designing and production team. Their endless efforts to recruit the best for this project, has resulted in the accomplishment of this book. They are a veteran in the field of academics and their pool of knowledge is as vast as their experience in printing. Their expertise and guidance has proved useful at every step. Their uncompromising quality standards have made this book an exceptional effort. Their encouragement from time to time has been an inspiration for everyone.

The publisher and the editorial board hope that this book will prove to be a valuable piece of knowledge for students, practitioners and scholars across the globe.

Index

Printed in the USA
CPSIA information can be obtained
at www.ICGtesting.com
LVHW010818221023
761596LV00005B/544

9 798888 360606